Springer Proceedings in Mathematics & Statistics

Volume 413

This book series features volumes composed of selected contributions from workshops and conferences in all areas of current research in mathematics and statistics, including data science, operations research and optimization. In addition to an overall evaluation of the interest, scientific quality, and timeliness of each proposal at the hands of the publisher, individual contributions are all refereed to the high quality standards of leading journals in the field. Thus, this series provides the research community with well-edited, authoritative reports on developments in the most exciting areas of mathematical and statistical research today.

Oscar Castillo · Uttam Kumar Bera ·
Dipak Kumar Jana
Editors

Applied Mathematics and Computational Intelligence

ICAMCI-2020, Tripura, India, December 23–24

Springer

Editors
Oscar Castillo
Research Chair of Graduate Studies
Tijuana Institute of Technology
Tijuana, Mexico

Uttam Kumar Bera
Department of Mathematics
National Institute of Technology Agartala
Agartala, Tripura, India

Dipak Kumar Jana
Department of Applied Sciences
Haldia Institute of Technology
Haldia, West Bengal, India

ISSN 2194-1009 ISSN 2194-1017 (electronic)
Springer Proceedings in Mathematics & Statistics
ISBN 978-981-19-8193-7 ISBN 978-981-19-8194-4 (eBook)
https://doi.org/10.1007/978-981-19-8194-4

Mathematics Subject Classification: 80A20, 03E72, 47N40, 03F55, 03B15, 35R60, 93C62, 97C30

This Springer imprint is published by the registered company Springer Nature Singapore Pte Ltd.
The registered company address is: 152 Beach Road, #21-01/04 Gateway East, Singapore 189721, Singapore

Committee

Chief Patron

Prof. Harish Kumar Sharma, Director, National Institute of Technology Agartala, India

Patron

Prof. Ardhendu Saha, National Institute of Technology Agartala, India

Organizing Committee

Prof. Debasish Bhattacharya, National Institute of Technology Agartala, India
Dr. Apu Kumar Saha, National Institute of Technology Agartala, India
Dr. Baby Bhattacharya, National Institute of Technology Agartala, India
Dr. Abhijit Baidya, National Institute of Technology Agartala, India
Dr. Susmita Roy, National Institute of Technology Agartala, India
Dr. Mantu Das, National Institute of Technology Agartala, India
Dr. Pinki Majumder, National Institute of Technology Agartala, India
Dr. Kalyani Debnath, National Institute of Technology Agartala, India
Dr. Sayanta Chakraborty, National Institute of Technology Agartala, India
Dr. Sudipa Choudhury, National Institute of Technology Agartala, India
Dr. Jayanta Debnath, National Institute of Technology Agartala, India
Dr. Jayasree Chakraborty, National Institute of Technology Agartala, India
Dr. Piyali Debnath, National Institute of Technology Agartala, India
Deepshikha Sarma, National Institute of Technology Agartala, India
Akash Singh, National Institute of Technology Agartala, India

Anirban Tarafdar, National Institute of Technology Agartala, India
Saptadeep Biswas, National Institute of Technology Agartala, India
Jagannath Nath, National Institute of Technology Agartala, India
Md. Mirazul Hoque, National Institute of Technology Agartala, India
Shilpa Samaddar, National Institute of Technology Agartala, India
Mousami Dhar, National Institute of Technology Agartala, India
Saroj Kumar Sahoo, National Institute of Technology Agartala, India
Sushmita Sharma, National Institute of Technology Agartala, India
Dr. Paritosh Bhattacharya (Organizing Chairman), National Institute of Technology Agartala, India
Dr. Uttam Kuma Bera (Organizing Secretary), National Institute of Technology Agartala, India

Technical Program Committee

Biswajit Sarkar, Yonsei University, South Korea
Prof. Oscar Castillo, Tijuana Institute of Technology, Mexico
Palash Dutta, Dibrugarh University, Assam, India
Indrajit Pal, Asian Institute of Technology, Thailand
Angeles Martinez Calomardo, University of Trieste, Italy
Amrit Das, VIT, India
Stojan Radenović, University of Belgrade, Serbia
Bijoy Krishna Debnath, Tezpur University, Assam, India
Gyu M. Lee, Pusan National University, Republic of Korea
Sapan Kumar Das, NIT Jamshedpur, India
Dipak Kumar Jana, Haldia Institute of Technology, India
Andreja Tepavcevic, University of Novi Sad, Serbia
Avishek Chakraborty, Narula Institute of Technology, Kolkata, India
Binod Chandra Tripathy, Tripura University, India
Guiwu Wei, Sichuan Normal University, P.R. China
Sanjay Kumar, G. B. Pant University of Agriculture and Technology, India
Tanmoy Som, IIT BHU, India
Shibaprasad Sen, Future Institute of Engineering and Management, India
B. K. Singh, NERIST, Arunachala Pradesh, India
Sabah M. Alturfi, University of Kerbala, Kerbala, Iraq
Kushal Pokhrel, Sikkim Manipal Institute of Technology, India
Mohmed A. Hassan, Ain Shams University, Cairo, Egypt
Sutapa Pramanik, Vidyasagar University, India
Shyamal Kumar Mondal, Vidyasagar University, India
Neha Bhardwaj, Amity University, India
Md. Izahuruddin, Aligarh Muslim University, U.P. India
Prashant Patel, Xavier College, Ahmedabad, India
M. Nasiruzzaman, University of Tabuk, Saudi Arabia

S. Sindu Devi, SRMIST, Ramapuram, Chennai, India
Nabin Sen, Haldia Institute of Technology, India
Sankar Prasad Mondal, Maulana Abul Kalam Azad University of Technology, India
Krit Somkantha, Udon-Thani Rajabhat University, Thailand
Chanchal Kundu, Rajiv Gandhi Institute of Petroleum Technology, India
Mrinmoy Majumder. NIT Agartala, India
Hemen Dutta, Gauhati University, India
S. H. Dong, Instituto Politécnico Nacional, Mexico
Ujjwal Laha, National Institute of Technology, Jamshedpur, India
Peng Zhu, Nanjing University of Science and Technology, China
Ahmed Ismail Ebada, Mansoura University, Mansoura, Egypt
Santosh Chapaneri, University of Mumbai, India
Utpal Nandi, Vidyasagar University, India
Satyabrata Maity, University of Calcutta, India
Kalipada Maity, Mugberia Gangadhar Mahavidyalaya, India
Anupam Mukherjee, BITS Pilani Goa, India
Sujit Samanta, National Institute of Technology, Raipur, India
U. K. Misra, Berhampur University, Odisha, India
Asish Bera, Edge Hill University, UK
Jagannath Samanta, Haldia Institute of Technology, India
Banibrata Bag, Haldia Institute of Technology, India
Ajitha Soundararaj, National Institute of Technology Tiruchirappalli, India
Pradipta Banerjee, Sidho-Kanho-Birsha University, India
Debanjan Nag, ICFAI University Tripura, India

Preface

The 1st International Conference on Applied Mathematics and Computational Intelligence (ICAMCI-2020) was held at the National Institute of Technology Agartala, Tripura, India, from December 23 to 24, 2020. National Institute of Technology Agartala is an institute of the national importance in India for the past several years. It has gained its reputation through its institutional dedication to teaching and research.

The primary topic of the conference is mathematics—which is considered the queen of science—has always been a discipline of interest not only to theoreticians but also to practitioners of all professions. In various areas of science, technology, economics, medicine as well as sociology, new mathematical principles, and models have been emerging and helping in new research and in drawing conclusions from practical data. During past few decades, there has been an enormous growth in applications of mathematics, which are multidisciplinary in nature. Some of the main topics of this conference are artificial intelligence, fuzzy logic, machine learning, and decision making, which have got more focus recently due to their various applications. With emerging computing facilities and speed, an enormous growth has happened nowadays in problem-solving area because of the complex algorithms and computer usage for long calculations, verifications, or generation of huge amount of data. Encompassing mathematics and computing, including various aspects of big data analytics, this conference is organized to share experience and new innovative ideas for futuristic needs from different disciplines to tackle new challenging problems together.

It is our sincere hope that the conference will help the participants in their research and training and open new avenues for work for those who are either starting their research or are looking for extending their area of research to a different area of current research in Applied Mathematics and Computational Intelligence.

In response to the call for papers for ICAMCI-2020, 173 papers were submitted for presentation and inclusion in the proceedings of the conference. The papers were evaluated and ranked on the basis of their significance, novelty, and technical quality by at least two reviewers per paper. After a careful blind refereeing process, 78 papers were selected for inclusion in the conference proceedings. The papers cover current research in applied mathematics, computational intelligence, fuzzy sets, intelligent systems, soft computing, machine learning, and natural language processing, image and video processing, computer network and security, cryptography, and data handling, rough set, fuzzy logic, operations research, optimization, and uncertainty theory.

Eminent personalities both from India and abroad (Mexico, South Korea, Thailand, etc.) have participated in the conference and delivered invited talks. The speakers from India are recognized leaders in government, industry, and academic institutions like IIT Madras, IIT Bombay, ISM (IIT) Dhanbad, NIT Durgapur, etc. All of them are involved in research dealing with the current issues of interest related to the theme of the conference. There were three keynote talks by Prof. Oscar Castillo (Tijuana Institute of Technology, Mexico), Prof. Biswajit Sarkar (Yonsei University, South Korea), and Dr. Indrajit Pal (AIT, Bangkok), along with four invited talks.

In addition to paper presentations of contributed lectures in specialized sessions, the main conference included special invited addresses by experts. After a thorough evaluation procedure, 20 research papers of exceptional quality were chosen for publishing in this book out of the 78 that were chosen for presentation at the ICAMCI-2020 conference.

A conference of this kind would not be possible to organize without the full support from different people across different committees. All logistics and general organizational aspects are looked after by the organizing committee members who spent their time and energy in making the conference a reality. We also thank all the technical program committee members and external reviewers for thoroughly reviewing the papers submitted for the conference and sending their constructive suggestions within the deadlines. Our hearty thanks to Springer for agreeing to publish the proceedings in the *Studies in Computational Intelligence* series.

We are grateful to the speakers, participants, referees, organizers, sponsors, and funding agencies for their support and help without which it would have been impossible to organize the conference, the workshops, and the tutorials. We owe our gratitude to the volunteers who work behind the scene tirelessly in taking care of the details in making this conference a success.

We are indebted to DST-SERB, India, and CSIR, India, for sponsoring the event. Their support has significantly helped in raising the profile of the conference.

Last but not least, our sincere thanks go to all authors who submitted their papers to ICAMCI-2020 and to all speakers and participants. We sincerely hope that the readers will find the proceedings stimulating and inspiring.

Tijuana, Mexico — Oscar Castillo
Haldia, India — Dipak Kumar Jana
Agartala, India — Uttam Kumar Bera

Message from the General Chair

It was a great pleasure for me to act as a General Chair in the 1st International Conference on Applied Mathematics and Computational Intelligence (ICAMCI-2020) was held at the National Institute of Technology Agartala, Tripura, from December 23 to December 24, 2020. This was a big challenge especially in the middle of COVID-19 pandemic, when conference participants could not attend due to the pandemic. However, still the participants attended the conference in online mode, and the conference was a great success.

Internationally renowned professionals in the mathematical sciences and related fields have been asked to present invited lectures and tutorials. These speakers' presentations covered a wide range of topics in engineering mathematics, such as fuzzy logic, intelligent computing, and uncertainty. In addition to paper presentations of contributed lectures in specialized sessions, the main conference included special invited addresses by experts. After a thorough evaluation procedure, 20 research papers of exceptional quality were chosen for publishing in this book out of the 78 that were chosen for presentation at the ICAMCI-2020 conference.

The main topic of the conference, Mathematics—the Queen of the Sciences, has always been a discipline of interest not only to theoreticians but also to practitioners of all professions. In various areas of science, technology, economics, medicine, or even sociology, new mathematical principles and models have been emerging and helping in new research and in drawing conclusions from practical data. During past few decades, there were an enormous growth in applications of mathematics which were multidisciplinary in nature. Some of the main topics of this conference artificial intelligence, fuzzy logic, machine learning, decision making, and others are the areas which got more focus recently due to the need of various fields of applications. With emerging computing facilities and speeds, an enormous growth has happened nowadays in problem-solving area because of the complex algorithms and computer usage for long calculations, verifications, or generation of huge amount of data. Encompassing mathematics and computing, including various aspects of big data analytics, this conference is organized to share experience and new innovative ideas for futuristic needs from different disciplines to tackle new challenging problems together.

I hope the conference exceeded the expectations of the attendees and achieved its goals, which included the sharing of concepts and recent research as well as the identification or resolution of new targets and challenges. Above all, I think that one of the conference's key goals—that young researchers and students would have new avenues to pursue in their future research—has been accomplished. I am hoping that this collection of the best articles will act as a tool for disseminating novel concepts and types of research.

Prof. (Dr.) Oscar Castillo
Professor of Computer Science
Graduate Division
Tijuana Institute of Technology
Tijuana, Mexico

Message from the Program Chair

It was a great pleasure for me to act as General Chair in the 1st International Conference on Applied Mathematics and Computational Intelligence (ICAMCI-2020) which was held at the National Institute of Technology Agartala, Tripura, from December 23 to December 24, 2020. Our main goal is to provide an opportunity to the participants to learn about contemporary research in applied mathematics and computing and exchange ideas among themselves and with experts present in the conference as the plenary as well as invited speakers. With this aim in mind, we carefully selected the invited speakers. It is our sincere hope that the conference will help the participants in their research and training and open new avenues for work for those who are either starting their research or are looking for extending their area of research to a different area of current research in Applied Mathematics and Computational Intelligence.

The conference was three keynote talks by Prof. Oscar Castillo (Tijuana Institute of Technology, Tijuana, Mexico), Prof. Biswajit Sarkar (Yonsei University, South Korea), and Dr. Indrajit Pal (AIT, Bangkok) and four invited talks.

After an initial call for papers, 173 papers were submitted for presentation at the conference. All submitted papers were sent to external referees, and after refereeing, 23 papers were recommended for publication for the conference proceedings that will be published by Springer in its Recent Advances in Intelligent Information Systems and Applied Mathematics.

We are grateful to the speakers, participants, referees, organizers, sponsors, and funding agencies for their support and help without which it would have been impossible to organize the conference, the workshops, and the tutorials. We owe our gratitude to the volunteers who work behind the scene tirelessly in taking care of the details in making this conference a success.

Dr. Uttam Kumar Bera
Department of Mathematics
National Institute of Technology Agartala
Jirania, Barjala, West Tripura
Tripura, India

About This Book

This book discusses the most recent breakthroughs in intelligent techniques such as fuzzy logic, neural networks, and optimization algorithms, as well as their application in the development of intelligent information systems using applied mathematics. The authors also discuss how these systems might be used in domains such as intelligent control and robotics, pattern recognition, medical diagnosis, time series prediction, and complicated problem optimization. It provides young researchers and students with new directions for their future study by exchanging fresh thoughts and finding new targets/problems. The book is aimed for math and computer science readers, especially professors and students interested in the theory and applications of intelligent systems in real-world applications.

The book publishes new developments and advances in the various areas of fuzzy, type-3 fuzzy, intuitionistic fuzzy, computational mathematics, block chain, creak analysis, supply chain, soft computing, fuzzy systems, hybrid intelligent systems, thermoelasticity, etc. It is intended to serve as a text for the researchers as well as for the students of engineering and applied sciences.

The text is composed of 18 chapters, each of which is developed independently of the other chapters. Chapter "Free Vibration Analysis of Generalized Thermoelastic Homogeneous Isotropic Plate with Two Temperature Theory" deals with the effect of two temperatures on the propagation of thermoelastic waves. Also introduced two-temperature generalized theory of thermoelasticity is used to investigate the propagation of thermoelastic waves in a homogeneous isotropic plate. Chapter "Analysis of Heat Transfer Coefficients and Pressure Drops in Surface Condenser with Different Baffle Spacings" provides an analysis of heat transfer coefficients and pressure drops in surface condenser with different baffle spacing which generalize many well-known CFD simulation results existed in the literature. Chapter "Integration of Big Data and Internet of Things (IoT): Opportunities, Security and Challenges" discusses various aspects and challenges of IoT, which may encounter while providing the service in respect to the volume, variety, and velocity. The author also suggested some techniques to resolve these problems for a better IoT-based system with integration of big data. Chapter "Intuitionistic Fuzzy Metrics and Its Application" develops the theory of distances between two nonlinear intuitionistic fuzzy sets (numbers) with respect to (pseudo) metrics. They presented some definitions of metric spaces and the formula of distance measure of two different sets via cumulative aggregated formula. In Chapter "Complex Structure of Number in Language Processing", author has tried to address the complex structure of numbers through language processing. Digital Newspaper using augmented reality attempts to integrate AR with the newspaper. The concept of audiovisual way of news representation in details through augmented reality (AR)-based android application has some novelty which has been explored in Chapter "Digital Newspaper Using Augmented Reality?". In Chapter "Nonlocal Fuzzy Solutions for Abstract Second Order Differential Equations", the concept of "Nonlocal Fuzzy Solutions for Abstract Second Order Differential Equations" is good, and the authors apply the concept of semigroup theory and suitable fixed point theorem which provides the new direction of these nonlocal fuzzy solutions. "Performance Assessment of Routing Protocols in Cognitive Radio Vehicular Ad Hoc Networks" deals with performance assessment of ADOV and DSR routing protocols in CR-VANET which has been discussed in Chapter "Performance Assessment of Routing Protocols in Cognitive Radio Vehicular Ad Hoc Networks". Infinite system of second-order differential equations in Banach space c0 is detailed in Chapter "Infinite System of Second Order Differential Equations in Banach Space c_0". Chapter "Fourth-Order Computations of Mixed Convection Heat Transfer Past a Flat Plate for Liquid Metals in Elliptical Cylindrical Coordinates" includes the fourth-order computations of mixed convection heat transfer past a flat plate for liquid metals in elliptical cylindrical coordinates. Chapter "Steady and Unsteady Solutions of Free Convective Micropolar Fluid Flow Near the Lower Stagnation Point of a Solid Sphere" provides the steady and unsteady solutions of free convective micropolar fluid flow near the lower stagnation point of a solid sphere, an algorithm to restore low-light images by enhancement of its inverted illumination map using dehazing which has been explored in Chapter "Low-Light Image Restoration Using Dehazing-Based Inverted Illumination Map Enhancement". Here the case of single zero Roulette to form a mathematical model for understanding the winning of bets has been discussed in Chapter "Mathematical Modeling of Probability and Profit of Single-Zero Roulette to Enhance Understanding of Bets". Chapter "Nonlinear Computational Crack Analysis of Flexural Deficit Carbon and Glass FRP Wrapped Beams" addressing the application of blockchain in food safety has been explored. A theoretical paper addressing the application of blockchain in food safety is Nonlinear Computational Crack Analysis of Flexural Deficit Carbon and Glass FRP Wrapped Beams. Here, a nonlinear finite element computational analysis of the RC beams has been retrofitted using carbon and glass FRP for flexural characteristic studies and comparison is performed with the test data which has been studied in Chapter "Economic Benefit Analysis by Integration of Different Comparative Methods for FACTS Devices". In Chapter "Optimal Pricing with Servicing Effort in Two Remanufacturing Scenarios of a Closed-Loop Supply Chain", the implementation of FACT devices with an eye to restore the quality of power supply has been studied. A classical optimization process under a closed-loop supply chain management with declining the emission of CO_2 considering carbon tax and selling price-dependent market demand has been studied in Chapter "A Review on Type-2 Fuzzy Systems in Robotics and Prospects for Type-3 Fuzzy". Some the definition of type-3 logic and their membership functions to enhance the quality of the chapter and proposing some prospect of type-3 fuzzy logic have been explored in Chapter "Optimization of Palm Oil Mill Effluent (POME) Solubilization Using Linguistic Fuzzy Logic and Machine Learning Techniques".

The editors believe that the choice of diversified advanced topics and their brief representation in this book would appeal to a wider section of research scholars, practicing engineers, and scientists.

Contents

About the Editors

Oscar Castillo holds the Doctor in Science degree (Doctor Habilitatus) in Computer Science from the Polish Academy of Sciences (with the Dissertation "Soft Computing and Fractal Theory for Intelligent and Manufacturing"). He is Professor of Computer Science in the Graduate Division, Tijuana Institute of Technology, Tijuana, Mexico. In addition, he is serving as Research Director of Computer Science and Head of the research group on Hybrid Fuzzy Intelligent Systems. Currently, he is President of Hispanic American Fuzzy Systems Association (HAFSA) and Past President of International Fuzzy Systems Association (IFSA). Prof. Castillo is also Chair of the Mexican Chapter of the Computational Intelligence Society (IEEE). He also belongs to the Technical Committee on Fuzzy Systems of IEEE and to the Task Force on "Extensions to Type-1 Fuzzy Systems". He is also Member of NAFIPS, IFSA, and IEEE. He belongs to the Mexican Research System (SNI Level 3). His research interests are in type-2 fuzzy logic, fuzzy control, neuro-fuzzy, and genetic-fuzzy hybrid approaches. He has published over 300 journal papers, 10 authored books, 50 edited books, 300 papers in conference proceedings, and more than 300 chapters in edited books, in total more than 1000 publications (according to Scopus) with h index of 82 and more than 23000 citations according to Google Scholar. He has been Guest Editor of several successful Special Issues in the past, like in the following journals: *Applied Soft Computing*, *Intelligent Systems*, *Information Sciences*, *Soft Computing*, *Non-Linear Studies*, *Fuzzy Sets and Systems*, *JAMRIS* and *Engineering Letters*. He is currently Associate Editor of the *Information Sciences Journal*, *Journal of Engineering Applications on Artificial Intelligence*, *International Journal of Fuzzy Systems*, *Journal of Complex Intelligent Systems*, *Granular Computing Journal*, and *Intelligent Systems Journal* (Wiley). He was Associate Editor of *Journal of Applied Soft Computing* and *IEEE Transactions on Fuzzy Systems*. He has been elected IFSA Fellow in 2015 and MICAI Fellow in 2016. Finally, he recently received the Recognition as Highly Cited Researcher in 2017 and 2018 by Clarivate Analytics and Web of Science.

Uttam Kumar Bera did his M.Sc. and Ph.D. from Vidyasagar University, West Bengal, in Applied Mathematics with specialization in Operations Research in the year of 2001 and 2008, respectively. He has been teaching mathematics both at undergraduate and postgraduate level. At present, he is working as Associate Professor in the Department of Mathematics, National Institute of Technology Agartala (An Institute of National Importance, Government of India). Beside the regular teaching and research activity, he is in the charge of Assoc. Dean (SW), Chairman, DPC, SW section, Coordinator, DST-FIST project and DST-STUTI project, etc. Dr. Bera is a hardworking and motivated person. Dr. Bera is Senior Member of IEEE Society and Life Member of Operational Research Society of India, Indian, Calcutta Mathematical Society, Kolkata, and IAPQR. Dr. Bera has published more than 90 papers in esteemed international/national journals like *IEEE Transaction of Fuzzy System*, *IEEE Transaction on Engineering Management*, *IEEE Access*, *Applied Soft Computing*, *Computers and Industrial Engineering*, *Computer in Industry*, *Neural Computing and Applications*, *APM*, *CAMWA*, *IJCSM*, *JOS*, *IJOR*, *IJAOM*, *ICJA*, *OPESARCH*, *JTRS*, etc. At present, five students are pursuing Ph.D. under him. He has received six research projects in the capacity of PI/Co-PI from different funding agency like DST, NBHM, CSIR, and UGC-NPIU and completed successfully. Recently, he recieved DST-STUTI project of 2 Crore 24 Lakhs for one year. He is Editorial Member of two international journals. He has successfully guided eight research scholars in this field of inventory control, transportation problem, and disaster management. He has awarded DST-International travel grand and conference grad from DST in the year of 2019.

Dipak Kumar Jana did his Ph.D. from Indian Institute of Engineering Science and Technology, Shibpur, and his postgraduation (M.Sc.) in Applied Mathematics with specialization in Operations Research from Vidyasagar University, West Bengal. He has qualified National Eligibility Test (NET-CSIR) for Junior Research Fellow (JRF) and GATE. He has been teaching mathematics both at undergraduate and postgraduate level. At present, he is working as Professor and Head of the Department in School of Applied Science and Humanities, Haldia Institute of Technology, Haldia, West Bengal. Dr. Jana is Member of Operational Research Society of India, Indian Science Congress Association, Calcutta Mathematical Society. Dr. Jana has published more than 98 papers in esteemed international journals like *Information Science*, *Jr. Cleaner Production*, *Applied Soft Computing*, *Computers and Industrial Engineering*, *Jr. of Separation and Purification*, *IJUFKS*, *IJCSM*, *JOS*, *IJOR*, *IJAOM*, *ICJA*, *OPESARCH*, *JTRS*, etc. He has also been the offer of eight books: *A Text Book of Engineering Operations Research*, *Gate Mathematics* (Vol.-1 and 2), *Advanced Engineering Mathematics*, *Advanced Numerical Methods*, *Basic Engineering Mathematics*. He has also reviewed several text and reference books of mathematics, and he is Editorial Member of five international journals. He has also been the Ph.D. guidance of several research scholars in this field. At present, eight students are pursuing Ph.D. under him, wherein three have already been awarded the degree.

Free Vibration Analysis of Generalized Thermoelastic Homogeneous Isotropic Plate with Two Temperature Theory

Ankit Bajpai and P. K. Sharma

Abstract Two-temperature generalized theory of thermoelasticity is used to investigate the propagation of thermoelastic waves in a homogeneous isotropic plate. The governing equations are converted to non-dimensional form, and the potential functions are introduced to these equations. The surfaces of the plate are assumed to be stress-free thermally insulated. Secular equations in closed form and isolated mathematical conditions for symmetric and skew-symmetric wave mode propagation are derived. The results for uncoupled, coupled, Lord–Shulman, and Green–Lindsay theories of thermoelasticity have been obtained as particular cases. The numerical solution is performed for copper material. The effect of two temperatures on the propagation of thermoelastic waves is presented graphically, and the results are compared theoretically as well as numerically with the one-temperature Lord and Shulman theory to show the effect of two temperatures.

Keywords Generalized thermoelasticity · Two temperature theory · Rayleigh–Lamb waves · Secular equations

1 Introduction

The theory of thermoelasticity describes the interdependency of thermal and mechanical fields. Duhamel and Neumann [1] formulated the governing equations for thermoelastic solid counting the effect of temperature on deformation. This theory is known as the uncoupled theory of thermoelasticity. But, there are two shortcomings in uncoupled thermoelasticity that are not compatible with physical observations. The first one is that the heat conduction equation has no elastic term. Hence, elastic

A. Bajpai (✉) · P. K. Sharma
Department of Mathematics and Scientific Computing, National Institute of Technology Hamirpur, Hamirpur 177005, India
e-mail: ankitbajpai@nith.ac.in

P. K. Sharma
e-mail: psharma@nith.ac.in

O. Castillo et al. (eds.), *Applied Mathematics and Computational Intelligence*, Springer Proceedings in Mathematics & Statistics 413,
https://doi.org/10.1007/978-981-19-8194-4_1

changes do not affect the temperature. The second one is that it has a parabolic-type heat conduction equation providing infinite speed for the deliverance of heat waves. Boit [2] terminated the first shortcoming by introducing an elastic term, called the mechanical coupling term, in the heat conduction equation, and the formulated theory is known as the coupled theory of thermoelasticity. Researchers [3, 4] proved experimentally that heat waves propagate with finite speed. The developments of generalized theories overcome the second shortcoming of the infinite speed of thermal waves. These theories shift the governing equation of heat transportation from parabolic type to wave type and hence allow the finite speed for the flow of thermal disturbances. Lord and Shulman [5] proposed the first generalized theory of thermoelasticity by introducing one relaxation time. Another theory of thermoelasticity involving two relaxation times has been proposed by Green and Lindsay [6]. Other models of generalized thermoelasticity are Green and Nagdhi [7–9], Hetnarski and Ignaczack [10, 11], Chandrasekhar and Tzou [12, 13], and three-phase lag model [14]. Applying these generalized thermoelastic models, a lot of work has been done by many researchers.

Chen and Gurtina [15] and Chen et al. [16] suggested two different temperatures for heat conduction in deformable bodies. These temperatures are termed as the conductive temperature and the thermodynamic temperature. They are separated by heat supply for time-independent situations, and if heat supply vanishes, both temperatures will be equal. But for problems that are time-dependent, two temperatures are in general distinct even though heat supply is zero. Warren and Chen [17] studied the thermal wave propagation in two-temperature theory. Thereafter, for many years, this theory was underestimated and ignored. But, in recent times, two-temperature theory has been noticed by many researchers. They obtained, further, advancement in two-temperature theory and explained its applications primarily describing the continuity of stress function as it is discontinuous for one-temperature theory [18]. Puri and Jordan [19] presented a complete description of the existence, structural stability, convergence, and spatial behavior of the two-temperature theory. Youssef [20] formulated a theory by introducing two temperatures in the generalized thermoelastic model of Lord and Shulman, known as two-temperature generalized thermoelasticity, and proved the uniqueness theorem for the homogeneous isotropic case. This theory separates the temperature produced by the thermal process (conductive temperature) and the mechanical process (thermodynamic temperature). Kumar and Mukhopadhyay [21] analyzed the response of thermal relaxation time on the propagation of plane waves in the domain of two-temperature thermoelasticity. Bala [22] described an extensive literature survey on two-temperature thermoelasticity. Sur and Kanoria [23] studied the problem of thermoelastic interactions in three-dimensional medium in the context of two-temperature theory. Recently, many authors have been analyzed the two-temperature effects on the several problems of generalized thermoelasticity [28–30].

Sharma et al. [24, 25] studied the propagations of thermoelastic waves, the generalized thermoelastic waves, and generalized viscothermoelastic waves in homogeneous isotropic plates. Kumar and Pratap [26] investigated the propagation of waves in homogeneous isotropic microstretch generalized thermoelastic plates. Kumar et al.

[27] analyzed the effects of two temperatures on the problem of a micropolar porous circular plate with three-phase lags.

In this study, Youssef's model of two-temperature generalized thermoelasticity has been used to investigate the wave propagation in generalized thermoelastic plates. Governing equations are solved by applying normal mode analysis. Secular equations are obtained for symmetric and skew-symmetric modes of wave propagation by applying stress-free insulated or isothermal and rigidly fixed insulated or isothermal boundary conditions. The effect of two temperatures is shown theoretically on the obtained secular equations for both modes of propagation. The secular equations obtained by Sharma et al. [24] are obtained as a special case by neglecting the two-temperature effects. For numerical computations, copper material is taken, and numerical results are presented graphically for symmetric and skew-symmetric modes in the context of Lord and Shulman's theory with two temperatures and without two temperatures. The effect of two temperatures is presented graphically. The results are compared with the existing literature in one-temperature theory.

2 Basic Equations

In the framework of two-temperature theory of generalized thermoelasticity [20], the governing equations for a linear, homogeneous, and isotropic thermoelastic material, when body forces and heat sources are absent, are given by

$$(\lambda + \mu)u_{j,ij} + \mu u_{i,jj} - \gamma\left(\theta + \nu\frac{\partial \theta}{\partial t}\right)_{,i} = \rho\,\frac{\partial^2 u_i}{\partial t^2},\ (i, j = 1, 2, 3), \tag{1}$$

$$K\phi_{,ii} = \rho C_E\left(\frac{\partial \theta}{\partial t} + \tau_0\frac{\partial^2 \theta}{\partial t^2}\right) + \gamma T_0\left(\frac{\partial u_{i,i}}{\partial t} + n_0\tau_0\frac{\partial^2 u_{i,i}}{\partial t^2}\right), \tag{2}$$

$$\sigma_{ij} = 2\mu e_{ij} + \lambda e_{kk}\delta_{ij} - \gamma\left(\theta + \nu\frac{\partial \theta}{\partial t}\right), \tag{3}$$

$$e_{ij} = \frac{1}{2}\left(u_{i,j} + u_{j,i}\right). \tag{4}$$

The relation between two temperatures is:

$$\phi - \theta = a\phi_{,ii} \quad \& \quad \theta = |T - T_0|,\quad with\frac{\theta}{T_0} \ll 1, \tag{5}$$

where a is the two-temperature parameter, u_i, where i = 1, 2, 3, denote the components of displacement vector, σ_{ij} are the components of stress tensor, e_{ij} are the components of strain tensor, $e_{kk} = u_{i,i}$ = dilatation $(i, k = 1, 2, 3)$, $\phi_0 = T_0$ is the reference temperature, n_0 is constant, λ and μ are Lame's parameters, K is the thermal conductivity, ρ and C_E are the density and specific heat at constant strain, respectively, and $\gamma = (3\lambda + 2\mu)\alpha_t$, α_t is the linear thermal expansion coefficient. Taking a = 0, one will get the results of one-temperature thermoelasticity.

The coupled theory of thermoelasticity, Lord and Shulman generalized thermoelasticity theory, and Green and Lindsay theory can be obtained by putting $\tau_0 = \nu = a = 0$, $n_0 = 1$, $\tau_0 > 0$, $\nu = 0$, $a = 0$, and $n_0 = 0$, $\tau_0 > 0$, $\nu > 0$, $a = 0$, respectively.

3 Formulation of the Problem

Consider an infinite elastic plate of finite thickness $2d$ which is homogeneous, isotropic, and thermally conducting with initial uniform temperature T_0. The middle plane of plate coincides with the x–y plane such that $-d \le z \le d$ and $-\infty < x, y < \infty$. The origin of the coordinate system is taken at any point of middle plane. The boundary surfaces $z = \pm d$ are considered to be stress-free thermally insulated.

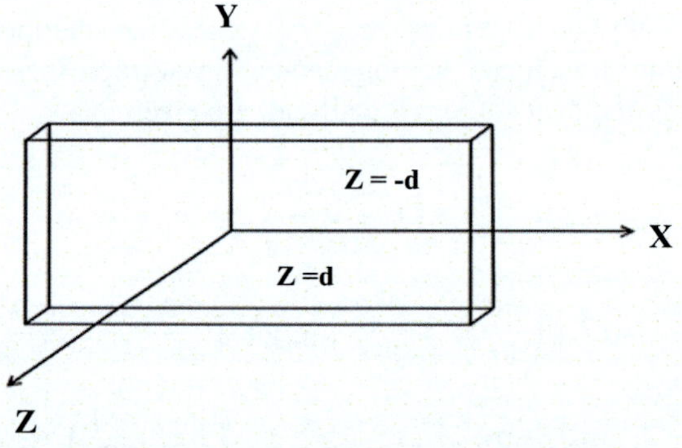

We consider the x–z plane as the plane of the incident and assume that the solutions are explicitly independent of y coordinate. Still, they depend implicitly so that the transverse component of displacement v does not vanish. Hence, the displacement components and temperatures are given by

$$u = u(x, z, t), \; v = 0, \; w = w(x, z, t), \; \phi = \phi(x, z, t) \;\&\; \theta = \theta(x, z, t)\,. \tag{6}$$

The governing field equations for the plate obtained from basic Eqs. (1)–(5) with the use of Eq. (6) are

$$(\lambda + 2\mu)\frac{\partial^2 u}{\partial x^2} + (\lambda + \mu)\frac{\partial^2 w}{\partial x \partial z} + \mu\frac{\partial^2 u}{\partial z^2} - \gamma\left(1 + \nu\frac{\partial}{\partial t}\right)\frac{\partial \theta}{\partial x} = \rho\frac{\partial^2 u}{\partial t^2}, \tag{7}$$

$$\mu\frac{\partial^2 w}{\partial x^2} + (\lambda+\mu)\frac{\partial^2 u}{\partial z\partial x} + (\lambda+2\mu)\frac{\partial^2 w}{\partial z^2} - \gamma\left(1+\nu\frac{\partial}{\partial t}\right)\frac{\partial\theta}{\partial z} = \rho\frac{\partial^2 w}{\partial t^2}, \tag{8}$$

$$\mu\left(\frac{\partial^2 v}{\partial x^2} + \frac{\partial^2 v}{\partial z^2}\right) = \rho\frac{\partial^2 v}{\partial t^2}, \tag{9}$$

$$K\left(\frac{\partial^2\phi}{\partial x^2} + \frac{\partial^2\phi}{\partial z^2}\right) = \rho C_E\left(\frac{\partial\theta}{\partial t} + \tau_0\frac{\partial^2\theta}{\partial t^2}\right) + \gamma T_0\left(\frac{\partial}{\partial t} + n_0\tau_0\frac{\partial^2}{\partial t^2}\right)\left(\frac{\partial u}{\partial x} + \frac{\partial w}{\partial z}\right). \tag{10}$$

Equation (9) is independent of the effect of two temperatures and has been already discussed by Sharma et al. [24]. Hence, it is not discussed afterward.

The constitutive relations are:

$$\sigma_{zz} = (\lambda+2\mu)\frac{\partial w}{\partial z} + \lambda\frac{\partial u}{\partial x} - \gamma\left(\theta+\nu\frac{\partial\theta}{\partial t}\right), \tag{11}$$

$$\sigma_{xz} = \mu\left(\frac{\partial u}{\partial z} + \frac{\partial w}{\partial x}\right). \tag{12}$$

Introducing non-dimensional quantities

$$x' = \frac{x\omega^*}{c_1},\ z' = \frac{z\omega^*}{c_1},\ u' = \frac{u\omega^* c_1\rho}{\gamma T_0},\ w = \frac{w\omega^* c_1\rho}{\gamma T_0},$$
$$t' = \omega^* t,\ \nu' = \omega^*\nu,\ \tau_0' = \omega^*\tau_0,\ \theta' = \frac{\theta}{T_0},\ \phi' = \frac{\phi}{T_0},$$
$$a' = \frac{a\omega^{*2}}{c_1^2},\ \delta^2 = \frac{\mu}{(\lambda+2\mu)}, \in = \frac{T_0\gamma^2}{\rho C_E(\lambda+2\mu)},$$

where

$$c_1^2 = \frac{(\lambda+2\mu)}{\rho}, \omega^* = \frac{C_E(\lambda+2\mu)}{K}.$$

The governing Eqs. (7), (8), and (10) in dimensional form are obtained as

$$\frac{\partial^2 u}{\partial x^2} + (1-\delta^2)\frac{\partial^2 w}{\partial x\partial z} + \delta^2\frac{\partial^2 u}{\partial z^2} - \frac{\partial}{\partial x}\left(\theta+\nu\frac{\partial\theta}{\partial t}\right) = \frac{\partial^2 u}{\partial t^2}, \tag{13}$$

$$\frac{\partial^2 w}{\partial z^2} + (1-\delta^2)\frac{\partial^2 u}{\partial x\partial z} + \delta^2\frac{\partial^2 w}{\partial x^2} - \frac{\partial}{\partial z}\left(\theta+\nu\frac{\partial\theta}{\partial t}\right) = \frac{\partial^2 w}{\partial t^2}, \tag{14}$$

$$\frac{\partial^2\phi}{\partial x^2} + \frac{\partial^2\phi}{\partial z^2} = \left(\frac{\partial\theta}{\partial t} + \tau_0\frac{\partial^2\theta}{\partial t^2}\right) + \in\left(\frac{\partial}{\partial t} + n_0\tau_0\frac{\partial^2}{\partial t^2}\right)\left(\frac{\partial u}{\partial x} + \frac{\partial w}{\partial z}\right). \tag{15}$$

Further, the mechanical boundary conditions in non-dimensional form at z = ± d are given as

$$\sigma_{zz} = \frac{\partial w}{\partial z} + (1 - 2\delta^2)\frac{\partial u}{\partial x} - \left(\theta + \nu\frac{\partial \theta}{\partial t}\right) = 0, \tag{16}$$

$$\sigma_{xz} = \delta^2\left(\frac{\partial u}{\partial z} + \frac{\partial w}{\partial x}\right) = 0. \tag{17}$$

The thermal boundary condition in non-dimensional form at z = ± d is given by

$$\frac{\partial \phi}{\partial z} = 0. \tag{18}$$

4 Solution of the Problem

In order to solve Eqs. (13) to (15), introduce potential functions Φ and Ψ by Helmholtz's decomposition theorem such that

$$u = \frac{\partial \Phi}{\partial x} + \frac{\partial \Psi}{\partial z}, \quad w = \frac{\partial \Phi}{\partial z} - \frac{\partial \Psi}{\partial x}. \tag{19}$$

And, take the solutions of the form

$$(\Phi, \ \Psi, \ \phi) = [f(z), \ g(z), \ h(z)]\exp[i\xi(x - ct)], \tag{20}$$

where $c = \omega/\xi$ is the phase velocity, ω is the frequency, and ξ is the wave number.

Inserting Eq. (19) in Eqs. (13), (14), and (15), further applying the solution defined by Eq. (20) in resulting equations, and then solving the resulting system of differential equations, the expressions for u, w, ϕ, and θ are obtained as

$$\begin{aligned} u(x, \ z, t) = {} & [i\xi(C_3\cos m_1 z + C_4 \sin m_1 z + C_5 \cos m_2 z + C_6 \sin m_2 z) \\ & -\beta C_7 \sin \beta z + \beta C_8 \cos \beta z]\exp[i\xi(x - ct)], \end{aligned} \tag{21}$$

$$\begin{aligned} w(x, z, t) = {} & [-m_1 C_3 \sin m_1 z + m_1 C_4 \cos m_1 z - m_2 C_5 \sin m_2 z + m_2 C_6 \cos m_2 z \\ & -i\xi(C_7 \cos \beta z + C_8 \sin \beta z)]\exp[i\xi(x - ct)], \end{aligned} \tag{22}$$

$$\phi(x, z, t) = \begin{bmatrix} \frac{1}{\alpha_1}(C_3 \cos m_1 z + C_4 \sin m_1 z) + \frac{1}{\alpha_2}(C_5 \cos m_2 z \\ +C_6 \sin m_2 z) \end{bmatrix} \exp[i\xi(x - ct)], \tag{23}$$

$$\theta(x,z,t)=\begin{bmatrix}\frac{\{1+a(\xi^2+m_1^2)\}}{\alpha_1}(C_3\cos m_1 z+C_4\sin m_1 z)\\ +\frac{\{1+a(\xi^2+m_2^2)\}}{\alpha_2}(C_5\cos m_2 z+C_6\sin m_2 z)\end{bmatrix}\exp[i\xi(x-ct)], \tag{24}$$

where

$$\alpha_j=\frac{-i\omega t_2\{1+ag_j\}}{\alpha^2-m_j^2},\quad g_j=\left(m_j^2+\xi^2\right),\quad (j=1,\ 2),$$

$$\alpha^2=\xi^2\left(c^2-1\right),\ \beta^2=\xi^2\left(\frac{c^2}{\delta^2}-1\right),$$

$$m_1^2=\frac{1}{2}\left(A+\sqrt{A^2-4B}\right),\ m_2^2=\frac{1}{2}\left(A-\sqrt{A^2-4B}\right),$$

$$A=\frac{2\xi^2-\omega^2\left(1+t_0-i\in\omega t_2t_3+a\left(\omega^2t_0+2i\xi^2\omega t_2t_3-2\xi^2t_0\right)\right)}{-1+a\omega^2(t_0-i\in\omega t_2t_3)},$$

$$B=\frac{-\xi^4-\omega^4t_0+\omega^2\xi^2\left(1+t_0-i\in\omega t_2t_3+a\left(-\omega^2t_0-i\in\xi^2\omega t_2t_3+\xi^2t_0\right)\right)}{-1+a\omega^2(t_0-i\in\omega t_2t_3)},$$

$$t_0=\tau_0+i\omega^{-1},\ t_2=\nu+i\omega^{-1},\ t_3=n_0\tau_0+i\omega^{-1}.$$

Terms corresponding to 'a' exhibit two-temperature thermoelasticity. If a = 0, one will get results for one-temperature thermoelasticity, which are in agreement with Sharma et al. [24].

5 Secular Equations

Using the above expressions (21)–(24) and applying boundary conditions (16), (17), and (18), a homogeneous system of linear equations in terms of unknowns C_i is obtained. The obtained homogeneous system of equations will be non-trivial iff the determinant of coefficient matrix vanishes. It gives the secular equation for stress-free thermally insulated boundary and is obtained as

$$\left(\frac{\tan m_1 d}{\tan\beta d}\right)^{\pm1}-\frac{m_1(\alpha^2-m_1^2)(1+a(\xi^2+m_2^2))}{m_2(\alpha^2-m_2^2)(1+a(\xi^2+m_1^2))}\left(\frac{\tan m_2 d}{\tan\beta d}\right)^{\pm1}$$
$$=\frac{4\xi^2\beta m_1(m_2^2-m_1^2)(1+a(\xi^2+\alpha^2))}{(\xi^2-\beta^2)^2(\alpha^2-m_2^2)(1+a(\xi^2+m_1^2))}. \tag{25}$$

In Eq. (25), the indices + 1 and –1 represent skew-symmetric and symmetric modes, respectively. Equation (25) is the characteristic equation for the modified guided two-temperature thermoelastic waves for propagating in the plate. We mention these waves as two-temperature thermoelastic plate waves instead of Lamb waves, whose characteristic was described by Lamb in 1971 for isotropic materials in elastokinetics. For one-temperature theory, the two-temperature parameter will be zero, i.e., $a = 0$. Then, the secular Eq. (25) reduces to the secular equation as described by Sharma et al. [24].

6 Numerical Results and Discussion

In order to describe theoretical results presented in above equations, numerical results are presented. Here, we have chosen copper material [18] for evaluation of numerical results and physical data which is as given below

$$\lambda = 7.76 \times 10^{10} Kg\, m^{-1} s^{-2},\ \ \mu = 3.86 \times 10^{10} Kg\, m^{-1} s^{-2},$$
$$\epsilon = 0.0168,\ \ \rho = 8954 Kg\, m^{-3},\ \ C_E = 383.1\, J\, Kg^{-1} K^{-1},$$
$$K = 386\, W\, m^{-1}\, K^{-1},\ \ \alpha_t = 1.78 \times 10^{-5} K^{-1},\ \ T_0 = 293K,$$
$$\tau_0 = 6.131 \times 10^{-13} s,\ \ \nu = 8.765 \times 10^{-13} s,\ \ d = 1.0.$$

The phase velocities of symmetric and skew-symmetric modes of wave propagation have been computed for various values of wave number from the dispersion relation (25) for stress-free thermally insulated boundary conditions for Lord and Shulman theory of thermoelasticity.

Figure 1 presents the variation of non-dimensional phase velocity with wave number for $n = 0, 1, 2$ in case of skew-symmetric modes of vibration. Curves with dark circles as markers correspond to two-temperature theory in both figures. The phase velocity in all the three wave modes has higher magnitude in two-temperature theory as compared to corresponding one-temperature theory of thermoelasticity. Similar behavior has been observed for other theories as well. Figure 2 presents the variation of phase velocity with wave number for $n = 0, 1, 2$ in case of symmetric modes of vibration. In case of one-temperature theory, the results become similar to existing literature in [24]. In symmetric modes of vibration, the phase velocity for all three wave modes in case of two-temperature thermoelasticity is having smaller magnitude than one-temperature theory. But in skew-symmetric modes of vibration, the phase velocity in two-temperature theory has a higher magnitude in comparison to one-temperature theory.

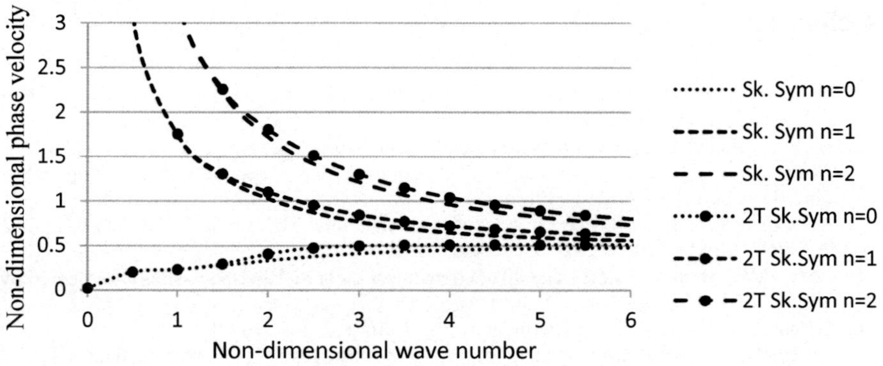

Fig. 1 Variation of phase velocity with wave number for skew symmetric mode in LS theory

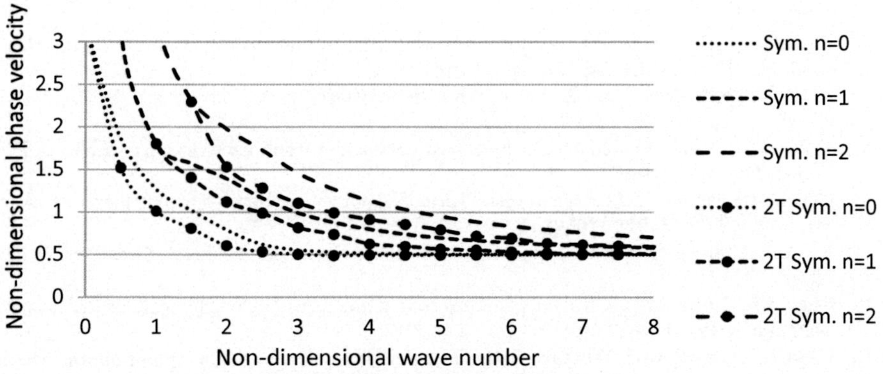

Fig. 2 Variation of phase velocity with wave number for symmetric mode in LS theory

7 Conclusions

Wave propagation in an infinite homogeneous isotropic plate of finite thickness '2d' in the context of two-temperature generalized thermoelastic model is investigated. Secular equations of stress-free thermally insulated plate surfaces for symmetric and skew-symmetric modes are derived. The phase velocity is obtained numerically from secular equations for Lord and Shulman theory in the stress-free thermally insulated boundary.

From the numerical results, one can observe that the two-temperature theory leads to an increase in the phase velocity in skew-symmetric modes and a decrease in the phase velocity in symmetric modes of vibration as compared to the one-temperature thermoelasticity.

References

1. Duhamel, J. M. C.: Second memorie surles. phenomenes thermo-mechaniques, J. L. Ecole Polytechn,15,1–57,1837.
2. Biot, M.: Thermoelasticity and irreversible thermodynamics. J. Appl. Phys. **27**, 240–253 (1956)
3. Ackerman, C.C., Bertman, B., Fairbank, H.A., Guyer, R.A.: Second sound in solid Helium. Phys. Rev. Lett. **16**, 789–791 (1966)
4. Ackerman, C.C., Overton, W.C.: Second sound in solid Helium-3. J. Phys. Rev. Lett. **22**, 764–766 (1969)
5. Lord, H.W., Shulman, Y.: A generalized dynamical theory of thermoelasticity. J. Mech. Phys. Solid. **15**, 299–309 (1967)
6. Green, A.E., Lindsay, K.A.: Thermoelasticity. J. Elast. **2**, 1–7 (1972)
7. Green, A.E., Nagdhi, P.M.: A reexamination of the basic results of thermomechanics. Proc. R. Soc. Lond. Ser. A **432**, 171–194 (1991)
8. Green, A.E., Nagdhi, P.M.: On undamped heat waves in an elastic solid. J. Therm. Stress **15**, 252–264 (1992)
9. Green, A.E., Naghdi, P.M.: Thermoelasticity without energy dissipation. J. Elast. **31**, 189–209 (1993)
10. Hetnarski, R.B., Ignaczak, J.: Soliton-like waves in a low temperature nonlinear thermoelastic solid. Int. J. Engng. Sci. **34**, 1767–1787 (1996)
11. Hetnarski, R.B., Ignaczak, J.: Generalized thermoelasticity. J. Therm. Stress **22**, 451–476 (1999)
12. Tzou, D.Y.: A unified field approach for heat conduction from macro to micro scales. J. Heat Transf. **117**, 8–16 (1995)
13. Chandrasekharaiah, D.S.: Hyperbolic Thermoelasticity: a review of recent literature. Appl. Mech. Rev. **51**, 705–729 (1998)
14. Roy Choudhuri, S.K.: On a thermoelastic three-phase-lag model. J. Therm. Stress **30**, 231–238 (2007)
15. Chen, P.J., Gurtin, M.E.: On a theory of heat conduction involving two temperatures. Z. Angew. Math. Phys. **19**, 614–627 (1968)
16. Chen, P.J., Gurtin, M.E., Willams, W.O.: On the thermodynamics of non-simple elastic material with two temperatures. Z. Angew. Math. Phys. **20**, 107–112 (1969)
17. Warren, W.E., Chen, P.J.: Wave propagation in the two temperature theory of thermoelasticity. Acta. Mech. **16**, 21–31 (1973)
18. Youssef, H.M., Al-Harby, A.H.: State-space approach of two-temperature generalized thermoelasticity of infinite body with a spherical cavity subjected to different types of thermal loading. Arch. Appl. Mech. **77**, 675–687 (2007)
19. Puri, P., Jordan, P.M.: On the propagation of harmonic plane waves under the two temperature theory. Int. J. Eng. Sci. **44**, 1113–1126 (2006)
20. Youssef, H.M.: Theory of two-temperature generalized thermoelasticity. IMA J. Appl. Math. **71**, 383–390 (2006)
21. Mukhopadhyay, S., Kumar, R.: Effects of thermal relaxation time on plane wave propagation under two-temperature thermoelasticity. Int. J. Eng. Sci. **48**, 128–139 (2010)
22. Bala, K.: A review on two-temperature thermoelasticity. IJMER **2**, 4224–4227 (2012)
23. Sur, A., Kanoria, M.: Three-Dimensional Thermoelastic Problem Under Two-Temperature Theory. Int. J. Comput. Methods **14**(3), 1–17 (2017)
24. Sharma, J.N., Singh, D., Kumar, R.: Generalized thermoelastic waves in homogeneous isotropic plates. J. Acoust. Soc. Am. **108**, 848–851 (2000)
25. Sharma, J.N., Singh, D., Kumar, R.: Propagation of generalized viscothermoelastic Rayleigh-Lamb waves in homogeneous isotropic plates. J. Therm. Stress **27**, 645–668 (2004)
26. Kumara, R., Partap, G.: Analysis of free vibrations for Rayleigh-Lamb waves in a microstretch thermoelastic plate with two relaxation times. J. Eng. Phys. Thermophys. **82**, 35–46 (2009)
27. Kumar, R., Miglani, A., Rani, R.: Generalized two temperatures thermoelasticity of micropolar porous circular plate with three phase lag model. J. Mech. **34**(6), 779–789 (2018)

28. Youssef, H.M., Bassiouny, E.: Two-temperature generalized thermopiezoelasticity for one dimensional problems. Comput. Methods Sci. Technol. **14**, 55–64 (2008)
29. Youssef, H.M., El-Bary, A.A.: Two-temperature generalized thermoelasticity with variable thermal conductivity. J. Therm. Stress **33**, 187–201 (2010)
30. Bera, M.B., Das, N.C., Lahiri, A.: Eigenvalue Approach to Two-temperature Generalized Thermoelastic Interactions in an Annular Disk. J. Therm. Stress **38**(11), 1308–1322 (2015)

Analysis of Heat Transfer Coefficients and Pressure Drops in Surface Condenser with Different Baffle Spacings

P. S. Prem Kumar, V. Surya Teja, S. Arunvinthan, and S. Nadaraja Pillai

Abstract Nowadays, heat exchange devices are becoming one of the essential components in complex engineering systems such as power plants and food processing industries. Especially in power plants, the surface condenser plays a crucial role in enhancing the thermal efficiency which works like shell and tube heat exchangers. Generally, the heat transfer coefficient and pressure drop of the surface condenser depend on baffle spacing. The baffle spacing significantly influences the heat transfer coefficient for the shell-side fluid. CFD simulations were carried out for different cases of single-pass shell and tube heat exchanger by varying the number of baffles at same operating conditions. The purpose of baffles is to support the tube bundle and directs the fluid to flow on the surface of tubes. In this study, four different cases by varying the shell diameter relative to baffle spacing and the number of baffles were considered to evaluate the heat transfer coefficients and pressure drops. It is observed that following the decrement in baffle spacing, the cross-flow area of shell-side region decreases; hence, there will be an increase in Reynolds number for shell-side fluid which results in enhanced heat transfer coefficients. Moreover, the segmental baffles are widely used in industrial applications where the purpose of enhancing heat transfer coefficient in surface condensers is to improve the condensation process of steam to liquid at faster rates.

Keywords Baffle spacing · CFD simulations · Heat transfer coefficient · Pressure drop · Surface condenser · Shell · Tube heat exchanger

P. S. Prem Kumar (✉)
Department of Aeronautical Engineering, Kumaraguru College of Technology, Coimbatore, India
e-mail: premkumar.ps.aeu@kct.ac.in

V. Surya Teja · S. Arunvinthan · S. Nadaraja Pillai
School of Mechanical Engineering, SASTRA University, Thanjavur, India
e-mail: nadarajapillai@mech.sastra.edu

O. Castillo et al. (eds.), *Applied Mathematics and Computational Intelligence*,
Springer Proceedings in Mathematics & Statistics 413,
https://doi.org/10.1007/978-981-19-8194-4_2

1 Introduction

Shell and tube heat exchangers (*S-T*'s) are widely utilized in major technical applications like thermal plants, pharmaceutical industries, oil refineries, food production industry, waste heat recovery plants, and air conditioning due to its lower production cost. *S-T*'s provide relatively more area and volume for the flowing fluids, thereby increase the heat transfer rates and further help reducing the difficulties associated with cleaning in the device. More than 35–45% of heat exchangers used in the industries are shell and tube heat exchangers due to their high adaptability and flexibility along with more convenient cleaning procedures in comparison with other types of heat exchangers. Shell and tube heat exchangers used in modern power plants are surface condenser type which plays a crucial role in converting the exhaust steam from the turbine into liquid condensate which is then utilized for boiling process. The major factor that is provided by *S-T* is more space to flow the exhaust steam which is at a low pressure, i.e., below than atmospheric pressure. As it is well known that the condenser maintains the low pressure to increase the turbine work output by varying the enthalpy at inlet and outlet of the turbine, it influences the thermal efficiency of the plant. The heat transfer mechanism in the surface condenser is the saturated steam and is condensed on the surface of tubes, and the circulated water gets heated inside the tubes. During condensation process, the temperature of shell-side fluid remains constant (phase change) because only the latent heat of steam releases which was taken by the water; so, to increase the heat transfer rate, the heat transfer coefficients must increase. This can be possible when the Reynolds number of shell-side fluid, i.e., steam is to increase by providing baffles. The major factors affecting the surface condenser performance are turbulence intensity, heat transfer coefficient, shell and tube flow rates, pressure drop, and fouling factors. Traditional *S-T*'s with segmental baffles are used in industries which shows a huge pressure drop. Reppich et al. [1] proposed a new technique for single-phase liquid and gaseous stream applications with segmental baffles in heat exchangers, and Soltan et al. [2] studied about optimizing the baffle spaces in condensers by developing the computer program to minimize the capital costs and operating costs. Wang et al. [3] studied about improving the performance of S-T's by replacing the segmental baffles with the helical baffles. Zhang et al. [4] conducted an experimental analysis on shell-side heat transfer for both standard segmental baffles and middle-overlapped helical baffles, in which they concluded that helical baffles show more effectiveness than segmental baffles. Wang et al. [5] demonstrated that inserting choke plates in between the two baffles reduces the outflow, but the shell-side performance was gradually reduced. The baffles used in their analysis are non-continuous helicoids.

Cao et al. [6] implemented new *S-T*'s with sextant helical baffles and reported that in sextant helical baffles, the leakage flow in triangle zone gets reduced and the radial velocity becomes more uniform. Similarly, Prithiviraj et al. [7, 8] together proposed a new technique for designing an *S-T* called distributed resistance approach; in this case, it is having multiple tubes in a single computational domain such that shell side of the heat exchanger was modeled by a coarse grid. Sha et al. [9] developed thermal hydraulic and multidimensional technique, in which shell-side designs consist of volumetric porosity and distributed resistance approaches. Stevanoivic et al. [10] completely performed a numerical approach for three-dimensional *S-T*'s by modeling baffles, shell and tube bundle with the porous media, in which these numerical results provide a better approach to experimental results. Bin Gao et al. [11] conducted experiments on *S-T*'s with the baffles having different helix angles such as 8°, 12°, 20°, 30°, and 40°, and the result showed that helix angle with 40° was better one. Sirous Zeyninejad Movassag et al. [12] replaced the segmental tube bundles with the helical bundles such that there will be a reduction in the pressure drop; hence, the pumping and operation costs are reduced. From the numerical studies, Farhad et al. [13] reported that with the same helix angle of 40° and for same operating conditions, the heat transfer area is decreasing with the increment of space between the baffles. But the heat transfer rates are getting increased with the increment of baffle spaces at the same pressure drop. In the same way, pressure gradient decreases with increase in the baffle spacing. You et al. [14] conducted numerical analysis on *S-T*'s with flower baffles by considering turbulence kinetic energy and its dissipation rate as well. The results showed that there is 15% error when compared with the experimental results. Wang et al. [15] conducted experiments on *S-T*'s with flowered baffles and original segmental baffles under same operating conditions and found that heat exchangers with flowered baffles are 20–30% better than segmental baffled heat exchangers. For calculating the pressure drop at shell side of *S-T*'s with standard segmental baffles, Edward Gaddis et al. [16] presented the complete procedure. Li et al. [17] found that the heat transfer coefficient increases with the increase in the baffle spacing, but it shows some effect on the distribution of heat transfer at each and every individual tube. Taborek et al. [18] suggested that baffle spacing of every shell and tube heat exchanger depends on shell diameter such that minimum spacing should be equal to 20% of shell diameter, whereas maximum spacing should not cross the shell diameter.

2 Design of Surface Condenser

A surface condenser is a device which is widely installed in thermal power plants to convert the exhaust steam coming out of the turbine into a liquid condensate which is then used for boiling purpose. The working of the surface condenser is like shell and tube heat exchanger. To study the effect of heat transfer coefficient and pressure drop in a condenser, the suitable design of *S-T*'s is to be chosen as per standards based on Kern [19] and Gaddis [20–22]. In most of the industrial applications, single-pass *S-T*'s are used because of the low production cost and easy manufacturing. In this study, a single-pass steam–water S-T is designed with a requirement to increase the turbine work output while maintaining the exhaust pressure below atmospheric as given in Table 1. The design of four different cases of the same dimensional condenser in 3-D view is depicted in Fig. 1. In general, the hot fluid in the condenser is steam which

Table 1 Data for design of surface condenser

Property	Unit	Shell-side fluid—steam value
T_{HI}	°C	45
T_{HO}	°C	45
P	kg/m^3	0.06555
C	KJ/kg K	1.9400
μ	kg/m-s	1.04606E-5
K	W/m–K	0.0199747
Property	Unit	Tube-side fluid—water value
T_{CI}	°C	25
T_{CO}	°C	35
P	kg/m^3	995.65
C	KJ/kg K	4.1800
μ	kg/m-s	0.0007973
K	W/m–K	0.6155
Fouling factor	m^2 K/W	0.0002

must be taken at the shell side of heat exchanger because it requires more numbers of the tube as the specific volume is very higher than water.

The other factor is that water has more fouling properties so that it should not be taken at the shell side because the shell-side cleaning is very difficult when compared with tube cleaning so that cold water should be taken at tube side which receives heat from the steam.

The number of baffles in the condenser depends on space between the two adjacent baffles. The number of baffles can be calculated by Eq. (1). To study the effect of heat transfer coefficient and pressure drop with different numbers of baffles in a condenser, the only possible way is to vary the baffle space because shell length of the condenser cannot be varied practically in any industrial applications. The four different cases are shown in Fig. 1.

$$N_B = \frac{\text{shelllength}}{\text{Bafflespace} + \text{Bafflethichness}} - 1 \tag{1}$$

3 CFD Simulation of Surface Condenser

A. *Physical Model*A commercially available CFD code (Fluent 15.0) has been

used to carry out the numerical simulations of the required geometries. A three-dimensional geometric model was designed in the Ansys Workbench followed by the appropriate meshing. The tube and shell materials considered in this

Fig. 1 Different baffle spacings for same dimensional condenser: case *B*—baffle space equals to 50% of shell diameter

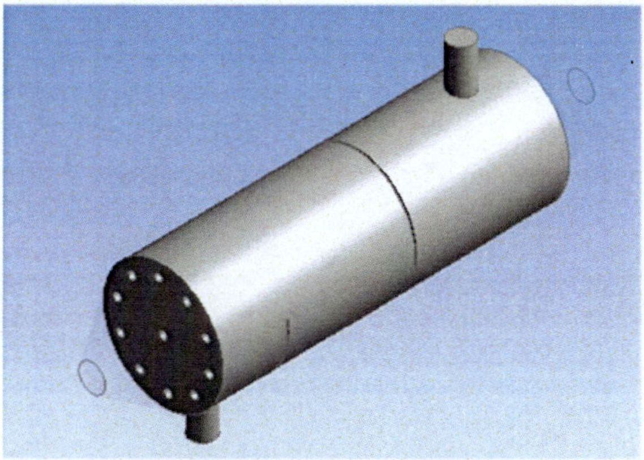

Case A – Baffle space equals to Shell diameter.

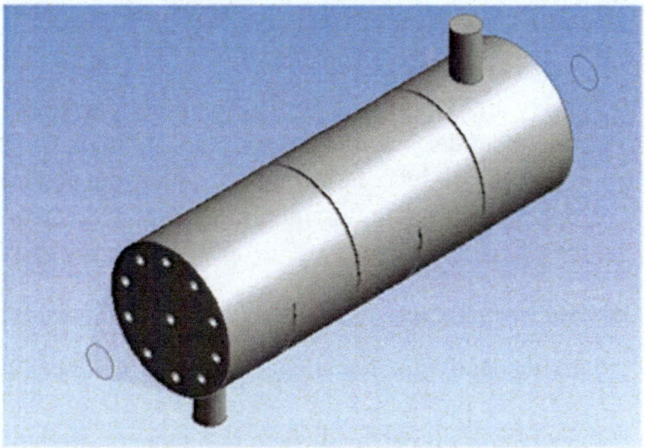

Case B – Baffle space equals to 50% of Shell diameter.

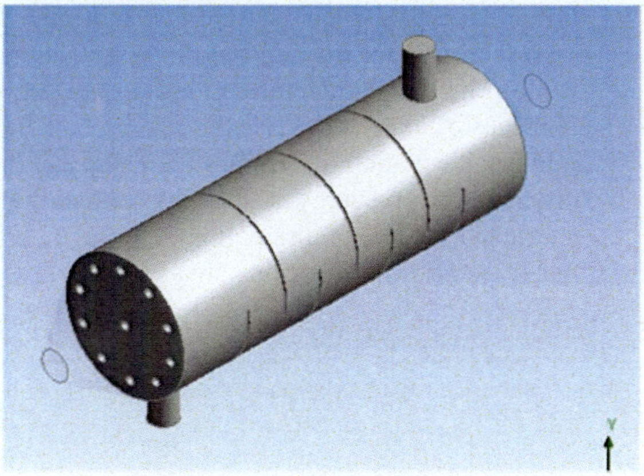

Case C – Baffle space equals to 25% of Shell diameter.

Fig. 1 (continued)

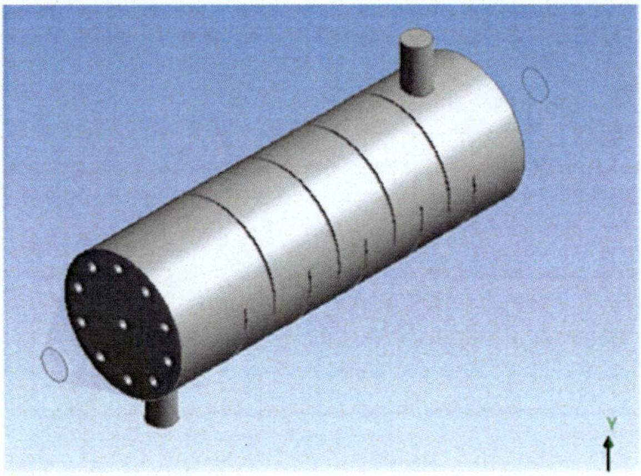

Case D – Baffle space equals to 20% of Shell diameter.

study are copper and steel respectively as available in the Fluent database. The inlet boundary conditions for both steam and cold water are set as flow openings, while the outlet conditions for condensate and hot water are set to be the pressure openings. The numerical simulations were solved to evaluate the heat transfer coefficients and pressure drop by using k-ε turbulence model.

B. *Governing Equations*

In this study, *k*-ε turbulence model is adopted. The governing equations used to solve the problem in the computational domain comprise continuity, momentum, energy, *k*, and ε. In surface condensers or shell and tube heat exchangers, the requirement of turbulence model is necessary because the turbulence intensity of fluid increases when the fluid flows through the tip of the baffles. The baffles are responsible for making fluid into zig-zag motion on the surface of the tube bundle. The energy transformation also happens between the shell and tube fluids by means of convection. *k*-ε model can be defined by two transport equations, of this one equation determines the energy in the turbulence and it is said to be turbulent kinetic energy which was denoted by *k*. The other transport equation determines the rate of dissipation of turbulent kinetic energy which was denoted by ε [23, 24]. The equations are as follows:

Turbulent kinetic energy (*k*):

$$\frac{\partial \rho k}{\partial t} + \frac{\partial (\rho k u_{i})}{\partial x_i} = \frac{\partial}{\partial x_i} \cdot \left[\frac{\mu_t}{\sigma_k} + \mu\right] \cdot \frac{\partial k}{\partial x_i} + Z_k \tag{2}$$

Turbulent dissipation energy (ε):

$$\frac{\partial \rho \varepsilon}{\partial t} + \frac{\partial (\rho u_i \varepsilon)}{\partial x_i} = \frac{\partial}{\partial x_i} \cdot \left[\frac{\mu_t}{\sigma_\varepsilon} + \mu\right] \cdot \frac{\partial \varepsilon}{\partial x_i} + Z_\varepsilon \tag{3}$$

$\underline{u}_t$ represents eddy viscosity.

$\underline{u}_i$ denotes velocity component in the corresponding direction.

Z_k and Z_ε represent the rate of deformation and buoyancy forces.

C. *Boundary Conditions*

- Shell-side inlet is set as flow opening of steam with different mass flow rates 1.389 kg/s, 2 kg/s, and 2.611 kg/s, respectively. The inlet temperature of the steam is considered as 318 K to maintain the constant temperature throughout the entire process.
- Tube-side inlet is set as inlet with a 1 m/s velocity of water at 298 K temperature.
- Both the outlets of shell and tubes are set as pressure openings, i.e., atmospheric pressure.

D .Test for Mesh Independency

Due to the complexity in the structure of a condenser, an unstructured grid with tetrahedral and triangular cells was considered in the computational domain. The grid independence test is required for any numerical study to ensure accuracy in the results. For the grid independence test, the surface condenser with four segmental baffles is meshed with different number of cells: 1.042, 1.140, 1.252, 1.306, 1.409 million. Numerical simulations were conducted for the same shell flow rate (1.389 kg/s) at various mesh densities to obtain the reliable solutions. The comparison between heat transfer rate and grid quantity is shown in Fig. 2, which results in the change in heat transfer rate of almost less than 3% in last three groups. Therefore, the change in the grid size has very little impact on heat transfer rates when the number of cells is near to 1.409 million. Consequently, the grid size giving the independent results is selected, and the total number of cells is adopted as 1.409 million cells. The convergence criteria for all required equations are to be assumed as 1×10^{-6} in residual monitors.

4 Results and Discussion

In this present study, four condensers with different baffle spacings and mass flow rates were numerically investigated to understand the heat transfer and flow field characteristics.

A. *Heat Transfer Coefficient*

Heat transfer coefficient versus mass flow rates of cases *A* to *D* are discussed in Fig. 3. It is clearly observed that the heat transfer coefficient increases with the decrease in baffle spacing as the Reynolds number of shell-side fluid depends on cross-flow area of a condenser. When the baffle spacing reduces with the increase in the number of baffles in the condenser, the cross-flow area reduces resulting in high turbulence intensity, i.e., enhanced local mixing of fluid results in the increment of heat transfer coefficients.

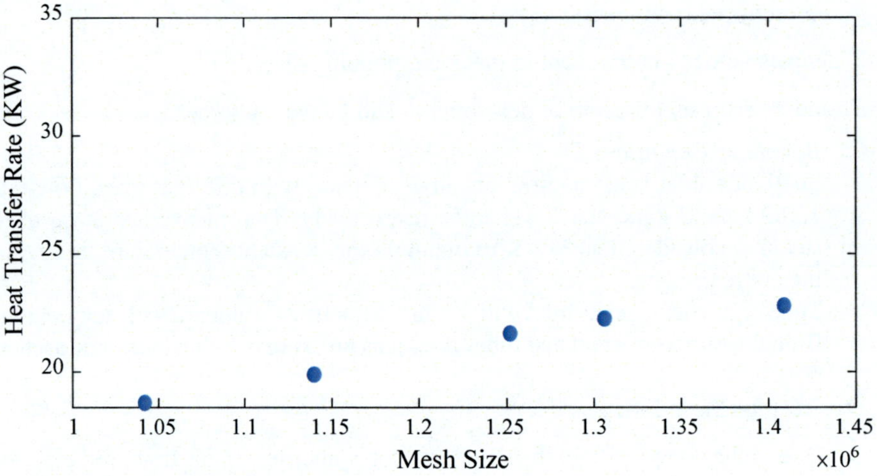

Fig. 2 Grid independence test of numerical simulation

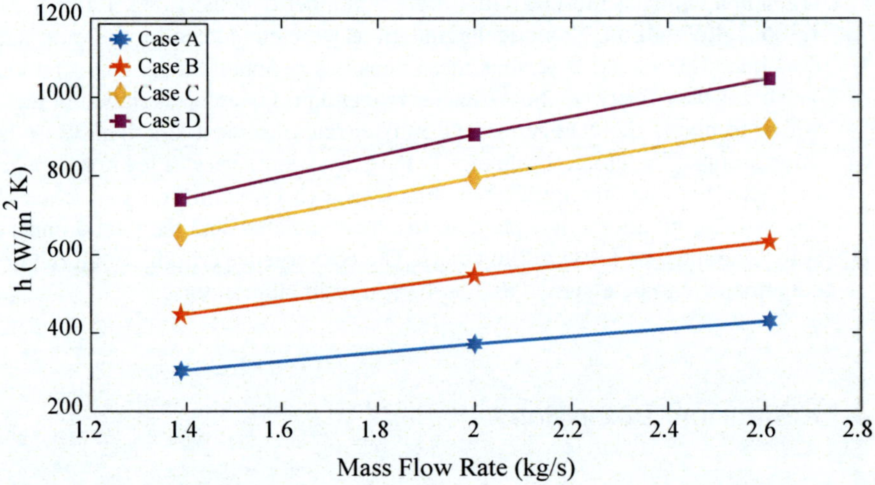

Fig. 3 Mass flow rate vs. heat transfer coefficient

For example, when comparing the heat transfer coefficients in cases *A* and *B*, *B* is nearly 31% greater than *A*. In the same way, the heat transfer coefficients in case C are further increased than case *B* by 32%. Case *D* has the highest heat transfer coefficients out of all the cases which is increased by 12% than case *C*.

B. *Pressure Drop*

Pressure drop versus mass flow rates of case *A–D* are discussed in Fig. 4. Pressure drop is considered as one of the important parameters in the design of

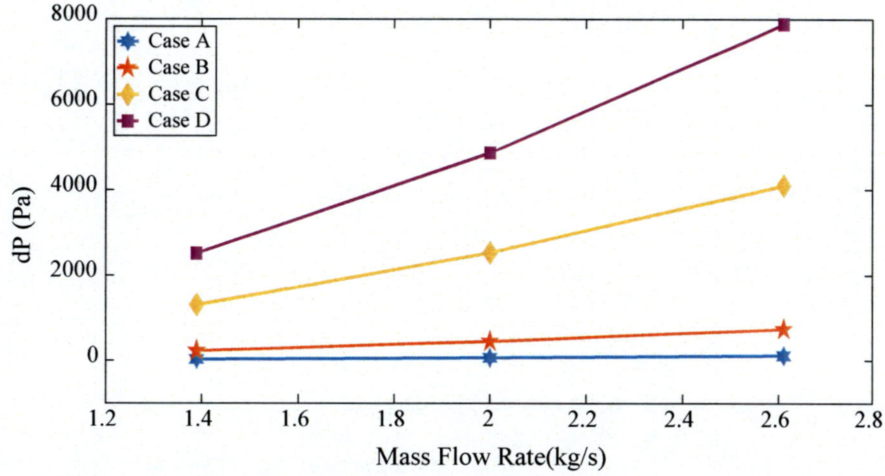

Fig. 4 Mass flow rate versus pressure drop

condensers as the pumping power used for shell fluid depends on it. Figure 4 clearly demonstrates that the pressure drop is reducing with the increase in the baffle spacing. As an example, the pressure drop in case *C* is 47% less than that of case *D*. The pressure drop reduced from case *C* to case *B* is 81% which is further reduced to 83% in case *A*. The reduction in pressure drops from case *D* to case *A* is attributed to the increase in the distance between the adjacent baffles and the corresponding decrement in the number of baffles, so that the flow area for shell fluid in the condenser is increasing; hence, there is little objection to the fluid flow.

C. *Heat Transfer Coefficient per Pressure Drop*

In order to obtain the optimum ranges between two characteristics like heat transfer coefficients and pressure drop, a considerable comparison should be done between them. Since the heat transfer coefficient and pressure drop are relying on each other, the comparison between these two characteristics' result shows that the baffle spacing can provide enhanced heat transfer within the acceptable range of pressure drop.

Figure 5 clearly shows that the condenser with longer baffle spacing shows higher heat transfer rates with the minimum pressure drop. For example, case B has heat transfer coefficient per pressure drop value less than 75% than case A which is further reduced to 73% in case *C*. The condenser with lower baffle spacing has smaller values as shown in Fig. 5 which has 83% lower than case *C*.

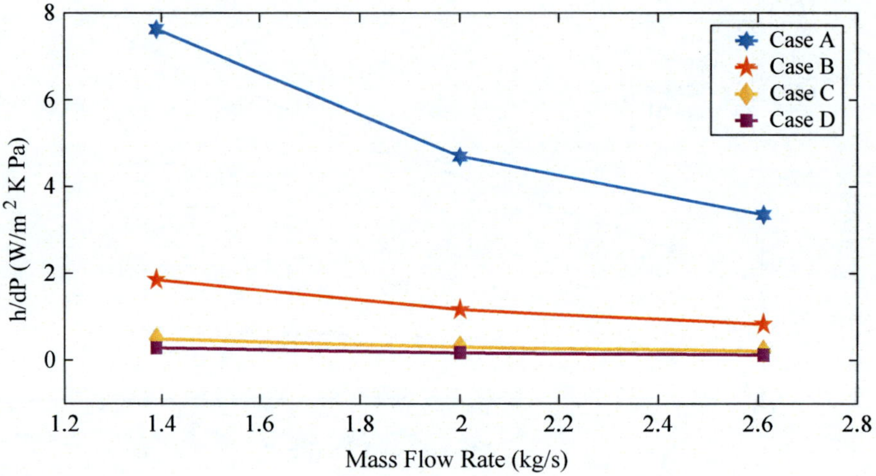

Fig. 5 Mass flow rate vs. heat transfer coefficient per pressure drop

5 Conclusion

In the present study, the numerical analysis of surface condenser with different baffle spacings was conducted to determine the effect of heat transfer coefficient and the corresponding pressure drop in the shell-side region of the surface condenser. CFD simulations were conducted for four condensers with same dimensions by varying the space between the baffles with the help of Fluent. The analysis on the condenser, the baffle spacing, and the number of baffles shows a significant effect on heat transfer area and turbulence intensity of shell-side fluid which alters the heat transfer coefficients and pressure drop. The main conclusions are as follows:

1. The increase in space between two adjacent baffles results in a decrement of pressure drop.
2. For the same working conditions and same mass flow rate, the heat transfer coefficients are predominantly increasing for the lower baffle spacing.
3. At the same pressure drop, lower baffle spacing has lower heat transfer coefficient as in condenser with nine baffles, whereas longer baffle spacing has higher heat transfer coefficient as in condenser with two baffles.

In the design of surface condensers, baffle space plays a crucial role. For instance, the design consideration towards the pressure drop and power required to pump the fluid into the shell, condenser with higher baffle spacing must be considered. If the priority is toward heat transfer coefficient, a condenser which provides higher turbulence intensity has to be chosen and this is possible with the more number of baffles and lower baffle spacing. In the future, the other prefix parameters like tube diameter, mass flow rate, baffle angle, and other such design aspects and its effect on heat transfer coefficients will be studied.

Acknowledgements This work was supported by "Research and Modernization fund, SASTRA Deemed to be University" grant number R&M/0035/SoME-008/2015-16. The authors thank SASTRA Deemed to be University for their financial assistance in performing the Analysis of Heat Transfer Coefficients and Pressure Drops in Surface Condenser with Different Baffle Spacings.

References

1. Reppich, M., Zagermann.: A new design method for segmental baffle heat exchangers. vol. I, no. 19, pp. 137–142, (1995)
2. Soltan, B.K., Saffar-Avval, M., Damangir, E.: Minimization of capital and operating costs of shell and tube condensers using optimum baffle spacing. Appl. Therm. Eng. **24**(17–18), 2801–2810 (2004)
3. Wang, G.D., Chen, G.D., Chen, Q.Y.: Review of improvements on shell and tube heat exchangers with helical baffles. Heat Transfer Eng. **31**(10), pp. 836–853 (2010)
4. Zhang, J.F., Li, B., Huang, W.J.: Experimental performance comparison of shell side heat transfer for shell and tube heat exchangers with middle-overlapped helical baffles and segmental baffles. Chemical Engineering and Sciences **64**, 1643–1653 (2009)
5. Wang, L., Luo, L.Q., Wang, Q.W.: Effect of inserting block plates on pressure drop and heat transfer in shell and tube heat exchangers with helical baffles". Journal of Engineering Thermodynamics **22**, 173–176 (2001)
6. Cao, X.:Theoretical analysis and experimental study of shell and tube heat exchangers with continuous helical baffles and overlapped helical baffles Doctoral dissertation. Shandong University. China
7. Prithiviraj, M., Andrews, M.J.: Three-dimensional numerical simulation of shell and tube heat exchangers. Part 1: foundation and fluid mechanics. Numerical Heat Transfer. Part- A Applications **33**, 799–816 (1998)
8. Prithiviraj,M., Andrews, M.M.J.: Three-dimensional numerical simulation of shell and tube heat exchangers. Part 2: Heat transfer. Numerical Heat Transfer. Part-A Appl **33**, pp. 817–28 (1998)
9. Sha, W.T., Yang, C.I., Kao, T.T., Cho, S.M.: Multidimensional numerical modelling of heat exchangers. J. Heat Transfer **104**, 417–425 (1982)
10. Stevanovic, Z., Ilic, G., Radojkovic, N., Vukic, M., Stefanovic, V., Vucˇkovic, G.: Design of shell and tube heat exchangers by using CFD technique – Part one: thermos hydraulic calculation. Facta Universitatis Series: Mechanical Engineering **8**, 1091–1105 (2001)
11. Bin Gao, Qincheng Bi, Zesen Nie, Wu Jiangbo, "Experimental study of effects of baffle helix angle on shell-side performance of shell-and-tube heat exchangers with discontinuous helical baffles. Experimental Thermal and Fluid Sci. **68**, pp.48–57 (2015)
12. Movassag, S.Z., Taher, F.N., Razmi, K., Azar, R.T.: Tube bundle replacement for segmental and helical shell and tube heat exchangers: performance comparison and fouling investigation on the shell side. Appl. Thermal Eng. **51**:1162–1169 (2013)
13. Taher, F.N., Movassag, S.Z., Razmi, K. and Azar, R.T.: Baffle space impact on the performance of helical baffle shell and tube heat exchangers. Appl. Thermal Eng. **44**:143–149 (2012)
14. You, Y., Fan, A., Huang, S., Liu, W.: Numerical modelling and experimental validation of heat transfer and flow resistance on the shell side of a s-t heat exchanger with flower baffles. Int. J. Heat Mass Transf. **55**:7561–7569 (2012)
15. Wang, Y., Liu, Z., Huang, S., Liu, W., Li, W.: Experimental investigation of shell-and-tube heat exchanger with a new type of baffles. Heat Mass Transf. **47**:833–839 (2011)
16. Edward, S., Gaddis, Gnielinski, V.: Pressure drop on the shell side of shell and the tube heat exchangers with segmental baffles. Chem. Eng. Process. **36**:149–159 (1997)

17. Li, H., Kottke, V.: Analysis of local shell-side heat and mass transfer in the shell and-tube heat exchanger with disc-and-doughnut baffles. Int. J. Heat Mass Transf. **42**:3509–3521 (1999)
18. Taborek, J.: Shell and tube heat exchangers: single phase flow, in heat exchanger. Design Handbook, Section 3.3, Hemisphere, New York (1982)
19. Kern, D.Q.: Process Heat Transfer. McGraw-Hill, New York, New York (NY) (1950)
20. Gaddis, D.: Standards of the tubular exchanger. Manufacturers Association, TEMA Inc, Tarrytown (NY) (2007)
21. Bayram, H., Sevilgen, G.: Numerical Investigation of the effect of variable baffle spacing on the thermal performance of a shell and tube heat exchanger. Energies **10**(1156) (2017)
22. Zebua, M.A., Ambarita, H.: A study on the effect of baffle spacing to the performance of a shell and tube heat exchanger. J. Phys. Conf. Ser. **1235**:012097 (2019)
23. Premkumar, P. S., Senthilkumar, C., Elangovan, S., Charavarthy Baskar, S.: Optimization of oil-cooler duct position for a pusher type turboprop. AIAA 2013–4331
24. Premkumar, P.S., Senthil Kumar, C.: Thrust loss optimisation in a turbo-prop aircraft using numerical simulation. Int. J. Appl. Eng. Res. 10(33):26623–26628, ISSN 0973–4562 (2015)

Integration of Big Data and Internet of Things (IoT): Opportunities, Security and Challenges

Suman Sekhar Sarangi, Debabrata Singh, Shrabanee Swagatika, and Nibedita Jagadev

Abstract IoT is upgrading and evolving due to its smart features and comfortless, whereas big data also helps storing and managing a huge amount of data effectively. In this paper we are trying to analyze the data created by IoT to provide an application and how effectively be managed by big data despite the data being structured, semi-structured or unstructured and while doing the management how cloud-based platform-as-a-service system (PaaS) is useful for a durable and cost-effective storage application. Later part of this paper we also discussed the various aspects and challenges of IoT may encounter while providing the services with respect to the volume, variety and velocity. We also suggested some techniques to resolve these problems for a better IoT-based system with integration of big data.

Keywords Big data · Cognitive · IoT · TLS · Hadoop · DWH · DL · HDFS · MRP

1 Introduction

IoT is a giant collection of physical objects or things connected through Internet that collects and exchanges data with each other [1]. As we know, people connect with each other with the help of social media over Internet, and here similarly, different things or objects connect with each other over Internet, hence the name as Internet of Things. The "thing" in IoT can be any objects such as automobile with a built-in sensor, fitness device that measures heart rate of a person, electronic appliances that

S. S. Sarangi · D. Singh (✉)
Department of CSIT, ITER, SOA Deemed to Be University, Bhubaneswar, Odisha, India
e-mail: debabratasingh@soa.ac.in

S. Swagatika · N. Jagadev
Deptartment of CSE, ITER, SOA Deemed to Be University, Bhubaneswar, Odisha, India
e-mail: shrabaneeswagatika@soa.ac.in

N. Jagadev
e-mail: nibeditajagadev@soa.ac.in

O. Castillo et al. (eds.), *Applied Mathematics and Computational Intelligence*, Springer Proceedings in Mathematics & Statistics 413,
https://doi.org/10.1007/978-981-19-8194-4_3

have unique identifiers (IPv6) [2] and are assigned with an IP address that collect data and transfer them over a network without human intervention. Machine to Machine communications is going to be the next generation of Internet revolution connecting huge number of devices on Internet [3].

Big data means large volume of complex data from different datasets which cannot be processed, analyzed and stored, using traditional data processing systems. Due to evolution and enhancement in technology, we are generating lots of data. Social media is one of the most important factors in the evolution of big data [4]. Data is generated through videos, images, etc., on a daily basis on an enormous amount. The role of big data in IoT is to process large amount of data generated by IoT devices on a real-time basis using different storage technologies [5].

1.1 Problem Statement

As the number of devices and data generated by IoT devices is increasing rapidly, therefore it is a challenging task to process, store and analyze tremendous amount of huge volume of complex data in scalable, cost-effective and distributive manner. So to tackle this situation, big data is used as a key to analyze the IoT-generated data from the connected devices. Its role is to process the large amount of data generated by IoT on a real time and store them using different technologies by dividing the complex data into small fragments and analyze them simultaneously.

1.2 Motivation

The primary motivation to work on this topic is finding loopholes of the IoT-generated data. The main concerns that accompany the IoT are various threats including the breach of privacy, overdependence on technology and the loss of jobs. Another report that took a gander at a great many associations from IoT gadgets present on big business systems found that over 40% of them do not scramble their traffic. Modern IoT gadgets helped us to connect virtually with different objects and adding a level of digital intelligence to devices. In this situation, it is very difficult for a programmer to control the traffic and collect the information. The greatest discovering was that 91.5% of information exchanges performed by IoT gadgets in corporate systems were decoded. To the extent gadgets go, 41% did not use Transport Layer Security (TLS) [6] by any stretch of the imagination, 41% utilized TLS just for certain associations, and just 18% utilized TLS encryption for all traffics. Gadgets that do not scramble their associations are powerless to different sorts of MitM assaults.

1.3 Challenges of IoT Data

In order to integrate big data with cloud, a lot of challenges may be encountered. Out of them, some are as follows:

- **Lack of updated software:** In order to use smart and compact IoT devices, we are not providing the better security as well as not providing the enough space to update the new versions of the software. The older versions of the product are also more prone to security theft.
- **Lack of secure communication:** To achieve high performance, the security is ignored which may be a challenge in long run. For a better IoT product, this part needs to be handled.
- **Authorized device:** Authenticity of IoT products is not verified in often which is another issue that needs to be rectified in order for this integration.
- **Botnet attacks:** The IoT devices are too often tested by various attacks to check their sustainability to which many devices fail to survive.
- **Cloud attack standards:** Due to the huge data associate with the IoT, some proper standard should be maintained to access and store the data in cloud. As cloud computing itself suffers from security issues, the devices which use the same technology will suffer from the same.
- **Security and privacy:** Security and privacy is an essential pillar of the Internet that is the most significant challenge for the IoT because any device without proper security and privacy is of no use in long aspect.

2 Related Works

Many researchers work on IoT as well as big data, and some also integrate both. For example, Kaur [7] integrates the big data with IoT, to find the critical data analysis and optimal solutions. In various systems of IoT devices, big data does not provide the sufficient information. So the integration improves the quality operations, i.e., education system, agriculture, environment protect, intelligent transportation, etc. Authors focused on multiple areas and proposed a new model, which can help to collect the data from various sources of IoT devices. Ge et al. [8] described about the big data for IoT. The disparity between the IoT domains has isolated the big data evolution. The mutual understanding helps the evolution of big data research in IoT. They discuss the similarity and differences among different big data technologies used in different IoT domains. Authors help to bridging the IoT and big data communities and selected a set of typical IoT domains and described the features in each domain. Four important aspects are also derived from the big data process.

Ahmed et al. [9] described about "the role of big data analytics in Internet of Things". Here, the researchers have examined the challenges associated with the successful deployment of IoT. Authors also focus on the recent advancement of big data analytics with IoT systems. They also manage the big data processing and different analytic solutions in the field of IoT environments. This paper identifies the opportunities resulting from the convergence of IoT, analytics and big data. Through the data analytics, the existence of big data solutions in IoT infrastructure can be taken care. Rehman et al. [10] describes about "The role of big data analytics in industrial Internet of Things". They investigated the recent big data analytics (BDA) technologies, algorithms and techniques that can lead to the development of intelligent IoT systems. Sassi et al. [11] described about a new architecture for Cognitive Internet of Things and big data. Here, the researchers proposed a new architecture that benefits computing mechanisms by combining the data warehouse (DWH) and data lake (DL) and defining a tool for heterogeneous data collection. They analyzed the existing technologies, tools and techniques from the related works and selected survey papers which allowed them to select appropriate technologies for their architecture. It could help future IoT researchers who would create architecture-based cognitive IoT concept that fits well with the scalability of the business requirements. They also intend to enhance the tool by considering new methods such as deep learning for extracting and recognizing the data from data sources that could be used to improve the data collection. Diene Bassirou [12] described about data management techniques for Internet of Things. This paper identifies the most relevant concepts of data management in IoT, surveys the current solutions proposed for IoT data management, discusses the most promising solutions and identifies relevant open research issues on the topic providing guidelines for further contributions. The researchers have provided new IoT data management methods such as middleware or architecture-oriented solutions to facilitate the integration of generated data, efficient storage and indexing methods of structured and unstructured data, as well as supports for NoSQL languages.

3 Classification of IoT

IoT is a giant collection of physical objects or things connected through Internet that collects and exchanges data with each other. As we know, people connect with each other with the help of social media over Internet, and here similarly, different things or objects connect with each other over Internet, hence the name as Internet of Things as depicted in Fig. 1.

3.1 Cognitive IoT

Cognitive systems are self-learning using different computing technologies like data mining, ML, natural language processing and human–computer interaction. CIoT is the amalgamation of various cognitive technologies which are collected from devices that are connected. The rise of presenting data to manage, process and store

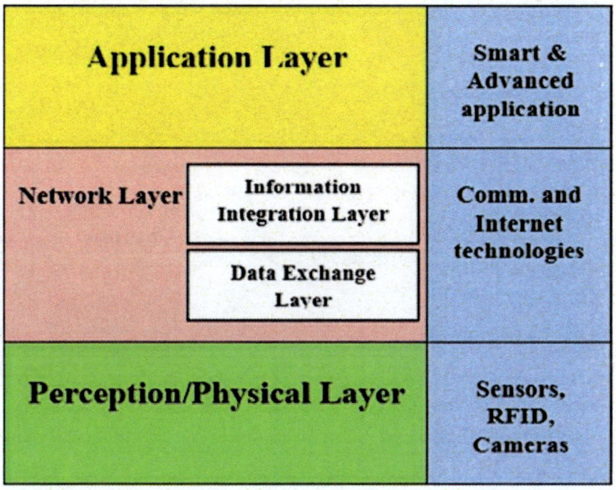

Fig. 1 Layers of IoT

information provides a route to diversified challenges of infrastructure. These various challenges are handled by IoT that are represented in factor awareness by developing a system that is smart enough to achieve requirements of the user. Therefore, Fig. 2 represents cognitive IoT in which data generated by web data (data is generated by various heterogeneous number of sources) and also IoT devices is observed through a middleware. Big data takes unstructured data, collected by IoT devices and organizes the to optimize their processes. IoT with big data process is depicted in Fig 3 and the features of hadoop are represented in Fig 4.

In the perception layer (lowest layer of the convulational architecture) user collects the useful information/data from things or the environment and sent to the smart

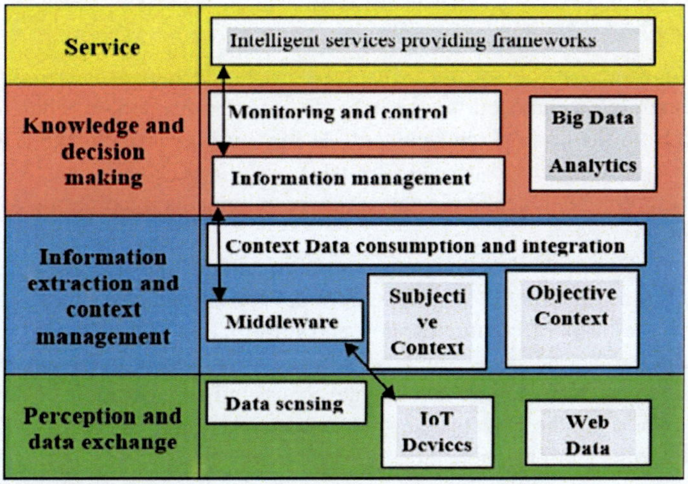

Fig. 2 Cognitive IoT architecture

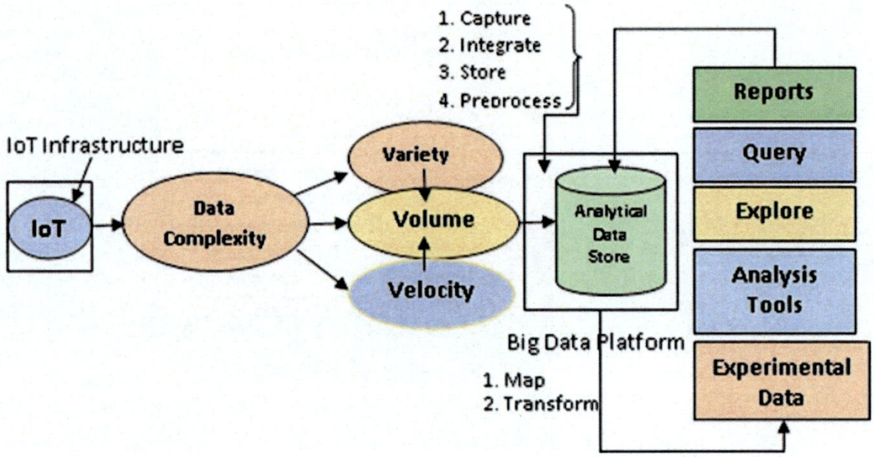

Fig. 3 IoT big data process

Fig. 4 Features of hadoop

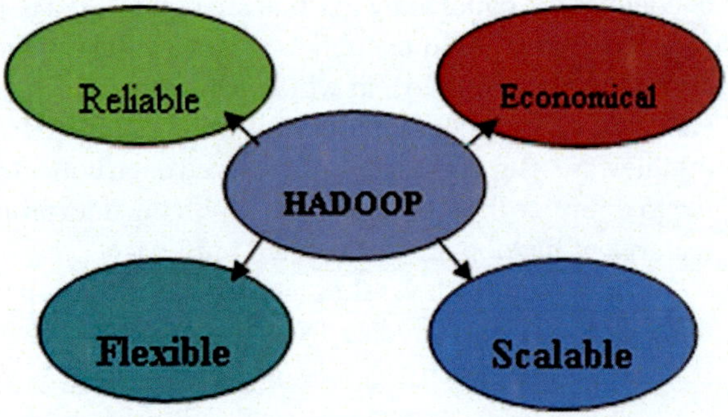

devices for store. In the second layer, information exchange, data reorganization and data extraction can occur by the help of context management. Then, the data is extracted from the context management. At the third layer, data lake (DL) occurs, transform to the dataset and stored in the data warehouse (DWH) through ETL. Then, data analysis can be done. At the service layer, it helps its maximum role and gives intelligent services.

Today, IoT is more an illustrative term of vision that everything should be connected to the Internet [13]. IoT is a connection of physical objects and virtual objects on the basis of communication and information technologies that provide opportunities to access data from any place. Physical world things are the objects that physically exist which we can operate, sense and connect to them, while things in virtual world are the objects that can be accessed, stored and processed. IoT uses sensors to collect information. Sensors are already used in day-to-day life which people are unaware of like smartphones that contain variety of sensors such as GPS, ambient light sensors, proximity sensors and torch screen sensors [14].

3.2 Applications of IoT

IoT is widely used in different contexts such as human body, homes, cities, industries and the global environment. The applications of IoT are:

- IoT in your home: Houses are equipped with smart devices like smart thermostats, lighting devices which can be remotely controlled via devices like smartphones and computers. For example, automatic watering system can be installed to keep plants in the garden alive and avoid wasting of water.
- IoT in human body: Here, IoT enables connectivity using sensors. For example, tracking activities and health status could improve person's daily life and also improve their future health.
- IoT in health care: IoT provides many benefits in the field of health care such as ability to keep track of patients closely to use the generated data which is analyzed afterward. Nowadays, hospitals use systems enabled with IoT to perform different tasks such as catalogue management, for medicines and different medical instruments [13].
- IoT in urban cities: In smart urban cities, different types of IoT detectors (or sensors) like smart streetlights help in reducing traffic, reduce air and noise pollution and improve sanitation [9].

3.3 Classification of Big Data

Big data means vast amount of enormous complex data which can be processed, analyzed and stored by using different types of data processing systems in a distributed manner. Now, you must be thinking big data as data that has huge memory size but that is not what exactly big data is [9]. Let us understand the concept of big data from an example. For example, if you try to link a document (of any type) of 200 MB to an email, then of course you would not be able to link the document to the email because the email system would not support the link (or attachment) which exceeds its size limit. Henceforth, the document in accordance to email can be termed as big data.

Let us discuss the concept of big data from a different real-life examples. Aeroplanes when they fly in the air, they transmit data continually to the air control systems which are built in the airports. The air control systems then use the data transmitted by the aeroplanes to track and detect the status, live location and progress of the aeroplanes in a distributed manner. As there will be multiple aeroplanes that would be transmitting data at the same time, a large volume of enormous data gets piled up at the air control systems within a short period of time. This enormous volume of data is termed as big data [15].

Big data is classified into three types:

- Structured data: The type of data which has a genuine (or proper) format associated with it is called as structured data. For example, data located in databases, CSV files, XLS, etc.
- Semi-structured data: The type of data that does not have any proper or genuine format associated with it is called as semi-structured data. For example, data that is present in log files doc files, emails, etc.
- Unstructured data: The type of data that does not have any type format associated with it is called as unstructured data. For example, graphic files, image files, mp3 files, mp4 files, etc.

Characteristics of Big Data:

Big data includes five V's. They are:

- Volume: The total amount of data that is processed (or generated) is termed as volume. The vast amount of data has become so huge that it cannot be stored and analyzed using traditional database technology.
- Velocity: The speed or rate at which the data is getting produced (or generated) is called as velocity. Nowadays, on a daily basis, amount of messages over social media, pictures and emails is increasing at a huge rate. Every second, data is increasing at an enormous speed. The technology of big data allows analyzing the data that is getting generated (or produced) without keeping it in the databases.
- Variety: Variety means the different types of data which are getting generated. Eighty percentage of data which are getting generated nowadays are unstructured that include graphic files such as images, videos. The technology of big data allows both structured data and unstructured data to analyze, process and store data at the same time.
- Value: The generated data's worth is referred as value. Having innumerable amount of data is of no use if it cannot be turned into valuable data. The most important aspect is to know and understand that the cost benefits of accumulating and examining the data are to make sure that the generated data can be tracked (or monitored). Here, we need to make sure that whatever data we generate, it will actually help us and has some value to it.
- Veracity: It refers to the extent to which how accurate, truthful and precise the generated data is and how much we can rely on that data. It is the uncertainties and inconsistencies in the data. Generating enormous amount of data is useless if quality and honesty (or trustworthiness) of the data are compromised, i.e., data is inaccurate. For example, a software bug causes data to be calculated incorrectly.

4 Impact of IoT on Big Data

As the number of devices and data generated by IoT devices are increasing rapidly, therefore it is a challenging task to process, store and analyze big data in scalable, cost-effective and distributed manner. IoT gives us access to data from millions of devices and servers through sensors, and they produce sets of data that need to be processed, stored and analyzed using big data. IoT helps in reducing costs and increasing revenue, but at the cost of generating enormous data [16].

Despite the fact that big data and IoT developed gradually and independently, both of them have become interconnected (or interrelated) over the course of time. A study by Gartner states that somewhere around 4.4 trillion GB of data will get generated by 2020 through IoT and more than 26 billion of sensors and devices will be connected to the Internet, and $300 billion revenue will be generated [17]. All these devices will generate and share huge amount of data in real time. When organizations use this data for the purpose of analyzing, IoT plays an important role as it acts as the major source for the data and this is where big data comes into picture. Big data is used as a process to analyze the data generated by IoT-connected devices. The role of big data in IoT is to analyze and process huge amount of data in a distributed manner and then store them using different big data technologies.

4.1 IoT Big Data Process

The process of IoT big data has four steps which are sequential in order:

i. Enormous amount of complex unstructured data gets generated by IoT devices that are collected in big data systems. This data generally depends on three V's of big data (volume, velocity and variety).
ii. Big data systems are a shared distributed type of database where the large amount of complex data gets stored in files of big data and then they are mapped and transformed into big data analysis.
iii. Then, the IoT-generated big data is analyzed using different big data analytic tools like MapReduce, Apache Hadoop, Spark, etc.
iv. Finally, generate the reports of the analyzed data.

4.2 How They Impact Each Other?

The impact of the huge amount of IoT-generated data is felt among all big data systems, as a result of which forcing the big data systems to upgrade their contemporary tools and technologies, to analyze and store data and then to take benefits of this newly generated data. Most of the organizations nowadays are acquiring the storage platform-as-a-service (PaaS) model in place of keeping their own infrastructure to store this huge data that will require development to handle the load of the big data.

Platform-as-a-service is a cloud-based solution which provides flexibility, scalability and an architecture that is used to store valuable data generated by IoT. The storage option of PAAS comprises public, private and hybrid models. If the organizations have sensitive data, best option is to use the private model to store the data; otherwise, public or hybrid model can be used to store the IoT-generated data. Since the data generated by IoT is unstructured, so big data needs fast analysis with large queries to process the IoT-generated data to perform rapid decisions. Therefore, the need of big data in IoT is enthralling. So, we can say that big data is the fuel which is used to run IoT.

4.3 How Unitedly They Are Beneficial to the Companies?

Both big data and IoT are useful to a diversity of data generated by IoT devices to examine and find out unrevealed data patterns and find out veiled corelations and unveil newly generated information. Since organizations could get benefit from the huge complex of IoT-generated big data, they handle (or manage) them to find exactly the way how data affect business. Henceforth, it helps companies to accomplish an improved and better understanding of the data, and as a result of which, companies take well-informed and efficient decisions.

Due to the rising demand of data storage, organizations prefer cloud storage of big data that lowers the cost of implementation for them. The future generation of e-healthcare systems will be improved by the combined features of the companies. The research of hypothesis-driven to data-driven will be guided by big data, and the various levels of connections between different sensors signals and already existing big data will be analyzed and controlled by IoT which will enable many different ways for discovering with a better understanding of the disease that leads to the development in the field of health care with the help of innovative solutions.

5 Massive Data Analysis from IoT Through Big Data

This explains the approach in which the huge enormous amount of data generated by IoT is processed and analyzed using different big data processing platforms.

5.1 Apache Hadoop

Written in Java programming language, Apache Hadoop is a big data processing platform that is used to process, analyze and store big data in a distributed manner in real time on a huge bunch of product hardware, and it is a software framework which is open source. Licensed under the second version of Apache (v2) license, it was developed by Google on the system of MapReduce and it applies the concepts of functional programming. Hadoop architecture consists of various components, among which the important components are MapReduce programming (MRP) and Hadoop Distributed File System (HDFS). Here, processing of data is done by MapReduce, and Hadoop Distributed File System is used for data storage. There were three major challenges to big data due to which Hadoop was introduced. They are:

1. Data storage: It was difficult to store the enormous amount of complex data using traditional systems as there was limited storage for a single system, and at the same time, data was increasing rapidly.
2. Storing heterogeneous data: There were different types of formats in which data was presented, i.e., structured, unstructured and semi-structured. Therefore, we need to have a type of system that can store different formats of data that is generated from different sources.
3. Processing speed of the data: The processing time of this huge amount of complex data is quite high.

5.1.1 Hadoop—An Emerging Solution

As of now, we have discussed three major setbacks or challenges faced by big data to process the IoT-generated data due to which we are using Hadoop as a solution to process, analyze and store the data. Since data storage is a major challenge faced, then, here in Hadoop, its component Hadoop Distributed File System (HDFS) provides a shared way for the storage of big data, where the data gets stored in the form of blocks in the data nodes and each data specifies size of each block. For example, if we have a data of size 1024 MB and HDFS is configured so that it will create 64 MB of blocks of data. Hence, HDFS will divide this data into 16 blocks as 1024/64 = 16 and store this data in the form of blocks among various data nodes. Data blocks get duplicated among different data nodes for providing fault tolerance during the time of storing the data. Among vertical and horizontal scaling, Hadoop follows the latter where new nodes can be added to the cluster of HDFS as per our own requirement, in spite of increasing hardware located at every node.

Another major setback of big data was to store heterogeneous data which gets resolved by using Hadoop because in HDFS, all kinds of data can be stored, i.e., structured, unstructured or semi-structured data. This is because HDFS follows a mechanism called write once and read many as a result of which any type of data can be written once and can be read multiple times to find perception. Last setback was the processing speed. For resolving this issue, the processing unit is moved to the data instead of moving data to the processing unit. In spite of moving data from various nodes to a single node (or master node) for processing, the logic of processing is directly sent to the nodes so that every node can process data in parallel where data is stored. At last, the final output generated by all the nodes gets merged, and they get sent to the client as a response.

5.1.2 Features of Hadoop

- Reliable: Hadoop is highly reliable. System is robust, i.e., when machines work together, if due to a technical failure one machine fails to work, then another will take the authority to work in a flexible, efficient and reliable manner.
- Flexible: Hadoop is highly flexible as it has the ability to work with all the different kinds of data, i.e., be it structured, unstructured or semi-structured, and it can process, analyze and store all the data.
- Scalable: By default, Hadoop can integrate smoothly with services that are cloud-based. If Hadoop is installed on cloud, then there is absolutely no need to worry about the scalability factor as we can get extra hardware and enlarge the system setup whenever required within a few minutes.
- Economical: The cost of possession of a project based on Hadoop is less as Hadoop uses commodity hardware. Maintaining Hadoop environment is easy and economical since Hadoop is open source due to which licensing is free.

5.2 Map Reduce

MapReduce is a big data processing platform that allows to perform dispense and parallel processing of datasets. It is also a programming framework that supports big data and various components of Apache Hadoop in order to improve its performance. For example, HDFS is replaced by Network File System for better security and high availability. MapReduce consists of two phases, i.e., map and reduce. In the map phase, data block is read and generated (or processed) to generate a key–value pair as midway outputs. Then, the reduce phase takes place where the output of the map phase acts as the input for the reduce phase where the reducer receives the key–value pair from different map jobs and reduces them to sets of small key–value pairs by aggregating them which happens to be the final output. However, complexity of MapReduce is more as compared to Apache Hadoop.

6 Conclusion

IoT is one of the largest sources of big data, which are processed or rendered without analytics power. Big data is introduced to process, analyze and store large amount of tremendous data generated by IoT device systems. So, IoT interacts with big data only when the large amount of complex data is needed to be processed and analyzed in higher frequency. In this paper, we first studied about IoT, its classification, applications and Cognitive IoT and the different challenges faced by IoT. To overcome this, we studied about big data, featured of big data and impact of IoT on big data and then analyzed the existing technologies, tools and equipment from the related works and several survey papers which allowed us to select the appropriate technology for the architecture. In addition, we have also surveyed the most important aspects of IoT and CIoT. We also identified the ways in how IoT and big data impact each other and how combinedly they are beneficial to the companies. We found that Apache Hadoop is the most reliable, economical, scalable and flexible among different big data processing applications and hence studied about Hadoop and how exactly Hadoop solves the problem of big data in processing the large amount of complex data in real time in a distributed manner. Finally, we conclude that the existing big data solutions in the IoT paradigm are still in their developing phase and the challenges related to them must be solved in the future.

References

1. Cecchinel, C., Jimenez, M., Mosser, S. and Riveill, M.: An architecture to support the collection of big data in the internet of things. In: 2014 IEEE World Congress on Services, pp. 442–449. IEEE (2014)
2. Montavont, N., Noel, T.: Handover management for mobile nodes in IPv6 networks. IEEE Commun. Mag. **40**(8), 38–43 (2002)
3. Hasan, M., Hossain, E., Niyato, D.: Random access for machine-to-machine communication in LTE-advanced networks: Issues and approaches. IEEE Commun. Mag. **51**(6), 86–93 (2013)
4. Marjani, M., Nasaruddin, F., Gani, A., Karim, A., Hashem, I.A.T., Siddiqa, A., Yaqoob, I.: Big IoT data analytics: architecture, opportunities, and open research challenges. IEEE Access **5**, 5247–5261 (2017)
5. Zhang, Q., Yang, L.T., Chen, Z., Li, P.: High-order possibilistic c-means algorithms based on tensor decompositions for big data in IoT. Inf. Fusion **39**, 72–80 (2018)
6. Salowey, J.: Transport layer security (TLS) session resumption without server-side state. Transport (2008)
7. Kaur, H., Kushwaha, A.S.:A review on integration of big data and IoT. In 2018 4th International Conference on Computing Sciences (ICCS), pp. 200–203. IEEE (2018)
8. Ge, M., Bangui, H., Buhnova, B.: Big data for internet of things: a survey. Futur. Gener. Comput. Syst. **87**, 601–614 (2018)
9. Ahmed, E., Yaqoob, I., Hashem, I.A.T., Khan, I., Ahmed, A.I.A., Imran, M., Vasilakos, A.V.: The role of big data analytics in Internet of Things. Comput. Netw. **129**, 459–471 (2017)
10. ur Rehman, M.H., Yaqoob, I., Salah, K., Imran, M., Jayaraman, P.P., Perera, C.: The role of big data analytics in industrial Internet of Things." Future Generation Computer Systems **99**:247–259 (2019)
11. Bhatt, C., Dey, N., Ashour, A.S (eds.).:Internet of things and big data technologies for next generation healthcare, pp. 978–3 (2017)
12. Sassi, M.S.H., Fourati, L.C. and Jedidi, F.G.: A new architecture for cognitive internet of things and big data. Proc. Comput. Sci. **159**(2019): 534–543
13. Diène, B., Rodrigues, J.J., Diallo, O., Ndoye, E.H.M., Korotaev, V.V.: Data management techniques for Internet of Things. Mech. Syst. Signal Process. **138**(2020): 106564 (2020)
14. Dey, N., Hassanien, A.E., Bhatt, C., Ashour, A. and Satapathy, S.C. eds.: Internet of things and big data analytics toward next-generation intelligence. Berlin: Springer (2018)

15. Chen, F., Deng, P., Wan, J., Zhang, D., Vasilakos, A.V., Rong, X.: Data mining for the internet of things: literature review and challenges. Int. J. Distrib. Sens. Netw. **11**(8), 431047 (2015)
16. Ding, G., Wang, L., Wu, Q.: Big data analytics in future internet of things. arXiv:1311.4112 (2013)
17. Hung, M.: Leading the IOT, Gartner insights on how to lead in a connected world. Gartner Res. 1–29 (2017)

Intuitionistic Fuzzy Metrics and Its Application

Kousik Bhattacharya and Sujit Kumar De

Abstract This article deals with qualitative difference between two intuitionistic fuzzy sets with the help of standard pseudo-metric and metric spaces. Some definitions over metric spaces, pseudo-metric spaces, intuitionistic fuzzy sets, indeterminacy and the formula of measuring metrices have been incorporated. Numerical illustrations, graphical illustrations, area of applications and ranking for decision-making are discussed to show the novelty of this article. Finally, conclusions and scope of future works are mentioned.

Keywords Metric distance · Pseudo-metric distance · Intuitionistic fuzzy set · Ranking

1 Introduction

In traditional set theory (classical), the idea of member and non-member of an element in a set was sudden, i.e. an element either belongs to a set or not belongs to the set. But there was no knowledge about the transition of an element from member to non-member of the set and vice-versa. Zadeh [1] has solved these ambiguities through his new invention, the fuzzy set theory. Since then, numerous research articles have been studied over the fuzzy set itself to explain the real-world phenomenon. Bellman and Zadeh [2] introduced a new concept of decision-making in a fuzzy environment. Piegat [3] gives us a new definition of fuzzy set. The concepts of dense fuzzy set studied by De and Beg [4, 5] to discuss the frequent learning effect of the fuzzy parameters. Analysing the behaviour of human thinking process, De [6] developed a new inexact set which is known as triangular dense fuzzy lock set and its new defuzzification method. After this invention many articles have been made by eminent researchers (Maity et al. [7, 8], De and Mahata [9], etc.) to control the individual or group decision-making problems on pollution sensitive inventory modelling. Baez-Sancheza et al. [10] discussed polygonal fuzzy sets and numbers

K. Bhattacharya (✉) · S. K. De
Department of Mathematics, Midnapore College (Autonomous), Midnapore 721101, India

O. Castillo et al. (eds.), *Applied Mathematics and Computational Intelligence*, Springer Proceedings in Mathematics & Statistics 413,
https://doi.org/10.1007/978-981-19-8194-4_4

extensively. Trapezoidal approximations of fuzzy numbers and their existence as well as uniqueness and continuity are exclusively discussed by Ban and Coroianu [11]. Chutia et al. [12] contribute to find membership function of a fuzzy number. De and Mahata [13] designed a fuzzy backorder model where demand rate is considered as cloudy inexactness. Decision-making in a bi-objective inventory problem was discussed by eminent researchers like De and Pal [14]. Mahanta et al. [15] made a new approach of fuzzy arithmetic without using alpha cuts. Mao et al. [16] extensively analysed about the relation between cloud aggregation operators and multiattribute group decision-making in interval valued hesitant fuzzy linguistic environment.

However, Atanassov [17, 18] introduces a new approach of fuzzy set, namely intuitionistic fuzzy sets in terms of membership and non-membership function. Some notable works over the EOQ models on fuzzy environments as well as IFS may be pointed out over here. A step order fuzzy approach is discussed by Das et al. [19]. Recently, Maity et al. [20] studied an intuitionistic dense fuzzy model where the learning–forgetting or agreement–disagreement is considered. De and Sana [21] discussed a stochastic demand model under aggregation with Bonferroni mean in intuitionistic fuzzy environment. Deli and Broumi [22] worked in neutrosophic soft matrices and NSM-decision-making. Recently, Kaur et al. [23] contributed to find relation between interval type intuitionistic trapezoidal fuzzy sets and decision-making with incomplete weight information. Liang and Wang [24] considered a linguistic intuitionistic cloudy fuzzy model with sentiment analysis in E-commerce. Xu [25] studied about intuitionistic fuzzy aggregation operators.

From the above discussion, it is observed that none of the researchers have been studied over the metric distances of intuitionistic fuzzy numbers. In this study, we develop the theory of distances between two nonlinear intuitionistic fuzzy sets (numbers) with respect to (pseudo) metrics. We give some definitions of metric spaces and the formula of distance measure of two different sets via cumulative aggregated formula. To show the novelty of this article, a numerical illustration has been analysed through the ranking of distances with the existing metrics.

2 Preliminaries

2.1 Here, We Shall Introduce Some Definitions Over Metric and Pseudo-Metric Spaces

Definition 2.1.1 Let A be a non-empty set. A function $d : A \times A \to \mathbb{R}$ is said to be a 'metric' or a distance function on A if it satisfies the following properties:

i. $d(x, y) \geq 0$ *for all* $x, y \in A$;
ii. $d(x, y) = 0$ *if and only if* $x = y$;
iii. $d(x, y) = d(y, x)$ *for all* $x, y \in A$;
iv. $d(x, z) = d(x, y) + d(y, z)$ *for all* $x, y, z \in A$.

Any non-empty set A together with a metric d defined on it is said to be a 'metric space'.

Definition 2.1.2 Let A be a non-empty set. A function $d^p : A \times A \to \mathbb{R}$ is said to be a 'pseudo-metric' on A if it satisfies the following properties:

i. $d^p(x, y) \geq 0$ *for all* $x, y \in A$;
ii. $x, y \in A$ and $x = y \Rightarrow d^p(x, y) = 0$;
iii. $d^p(x, y) = d^p(y, x)$ *for all* $x, y \in A$;
iv. $d^p(x, z) = d^p(x, y) + d^p(y, z)$ *for all* $x, y, z \in A$.

Any non-empty set A together with a pseudo-metric d^p defined on it is said to be a 'pseudo-metric space'.

Definition 2.1.3 [17] An intuitionistic fuzzy set A defined in the universe of discourse X is given by $A = \{x, \mu_A(x), \nu_A(x) | x \in X\}$, where $\mu_A, \nu_A : X \to [0, 1]$ denote the degree of membership and non-membership of x in A , respectively, satisfying the condition $0 \leq \mu_A(x) + \nu_A(x) \leq 1$. The indeterminacy degree $\psi_A(x) = 1 - \mu_A(x) - \nu_A(x)$ expresses the lack of knowledge of whether x belongs to A or not, also $0 \leq \psi_A(x) \leq 1$ for $x \in X$. An intuitionistic fuzzy number $\omega = (\mu_\omega, \nu_\omega)$ is an ordered pair which satisfies the conditions: $0 \leq \mu_\omega \leq 1$, and $0 \leq \mu_\omega + \nu_\omega \leq 1$, where and ν_ω are called membership degree and non-membership degree, respectively.

2.2 *Pseudo-Metrics in Intuitionistic Fuzzy Set*

Let us consider F be the set of all intuitionistic fuzzy numbers and each $\omega = (\mu_\omega, \nu_\omega) \in F$ is called a point of F. However, $\omega = (\mu_\omega, \nu_\omega, \psi_\omega)$ only has two degrees of freedom because $\mu_\omega + \nu_\omega + \psi_\omega \equiv 1$. So, we observe such a system by keeping one variable constant when the other variable is changing.

Definition 2.2.1 [25] Given two intuitionistic fuzzy numbers (IFN) ρ and σ,
$\rho \cap \sigma = \left(min\left(\mu_\rho, \mu_\sigma\right), max\left(\nu_\rho, \nu_\sigma\right)\right)$ and $\rho \cup \sigma = \left(max\left(\mu_\rho, \mu_\sigma\right), min\left(\nu_\rho, \nu_\sigma\right)\right)$.

Lemma 1 If $\left(\mu_\rho - \mu_\sigma\right)\left(\nu_\rho - \nu_\sigma\right) \geq 0$, $\rho \cap \sigma \in F$ and $\rho \cup \sigma \in F$.

Proof

$$\mu_{\rho\cap\sigma} + \nu_{\rho\cap\sigma} = \mu_\rho + \nu_\sigma \leq \mu_\sigma + \nu_\sigma \leq 1,$$

$$\mu_{\rho\cup\sigma} + \nu_{\rho\cup\sigma} = \mu_\sigma + \nu_\rho \leq \mu_\sigma + \nu_\sigma \leq 1, if \, \mu_\rho \leq \mu_\sigma;$$

$$\mu_{\rho\cap\sigma} + \nu_{\rho\cap\sigma} = \mu_\sigma + \nu_\rho < \mu_\rho + \nu_\rho \le 1,$$

$$\mu_{\rho\cup\sigma} + \nu_{\rho\cup\sigma} = \mu_\rho + \nu_\sigma \langle \mu_\rho + \nu_\rho \le 1, if \mu_\rho \rangle \mu_\sigma;$$

Definition 2.2.2 An ordered pair $\left(F, d_\psi^P\right)$ is called the intuitionistic fuzzy indeterminacy pseudo-metric space on F F, where $d_\psi^P : F^2 \to \mathbb{R}$ $d_\psi^P : F^2 \to \mathbb{R}$ is the indeterminacy pseudo-metric, for any $\rho, \sigma \in F$, $d_\psi^P(\rho,\sigma) = \frac{|\psi_\rho^2 - \psi_\sigma^2|}{2}$ $d_\psi^P(\rho,\sigma) = \frac{|\psi_\rho^2 - \psi_\sigma^2|}{2}$.

It is easy to verify that d_ψ^P d_ψ^P satisfies the properties of pseudo-metric. Similarly, we can define the intuitionistic fuzzy membership pseudo-metric space on F F, where $d_\mu^P : F^2 \to \mathbb{R}$ $d_\mu^P : F^2 \to \mathbb{R}$ is the membership pseudo-metric, for any $\rho, \sigma \in F$, $d_\mu^P(\rho,\sigma) = \frac{|\mu_\rho^2 - \mu_\sigma^2|}{2}$ $d_\mu^P(\rho,\sigma) = \frac{|\mu_\rho^2 - \mu_\sigma^2|}{2}$ and the intuitionistic fuzzy non-membership pseudo-metric space on F F, where $d_\nu^P : F^2 \to \mathbb{R}$ $d_\nu^P : F^2 \to \mathbb{R}$ is the non-membership pseudo-metric, for any $\rho, \sigma \in F$, $d_\nu^P(\rho,\sigma) = \frac{|\nu_\rho^2 - \nu_\sigma^2|}{2}$.

It is also easy to verify that d_μ^P and d_ν^P satisfies the properties of pseudo-metric.

Definition 2.2.3 An ordered pair (F, d_ψ) is called the intuitionistic fuzzy indeterminacy metric space on F, where $d_\psi : F^2 \to \mathbb{R}$ is the indeterminacy metric, for any $\rho, \sigma \in F$, $d_\psi(\rho,\sigma) = |\psi_\rho - \psi_\sigma|$.

It is easy to verify that d_ψ satisfies the properties of metric. Similarly, we can define the intuitionistic fuzzy membership metric space on F, where $d_\mu : F^2 \to \mathbb{R}$ is the membership metric, for any $\rho, \sigma \in F$, $d_\mu(\rho,\sigma) = |\mu_\rho - \mu_\sigma|$ and the intuitionistic fuzzy non-membership metric space on F , where $d_\nu : F^2 \to \mathbb{R}$ is the non-membership metric, for any $\rho, \sigma \in F$, $d_\nu(\rho,\sigma) = |\nu_\rho - \nu_\sigma|$.

It is also easy to verify that d_μ and d_ν satisfies the properties of metric.

Lemma 2 If $\alpha, \beta \in F$, $d_\Sigma(\alpha,\beta) = max\left(d_\mu(\alpha,\beta), d_\nu(\alpha,\beta), d_\psi(\alpha,\beta)\right)$ satisfies the four properties of pseudo-metric as well as metric.

Lemma 3 Let $\tilde{P}$ and $\tilde{Q}$ be two intuitionistic fuzzy sets defined over the interval $[L, R]$, then the summative metric distance (summative pseudo-metric distance) between $\tilde{P}$ and $\tilde{Q}$ is denoted by $d\left(\tilde{P}, \tilde{Q}\right)$ (for pseudo-metric $d_p\left(\tilde{P}, \tilde{Q}\right)$ and defined as.

$$d\left(\tilde{P}, \tilde{Q}\right) = \frac{1}{R-L} \int_L^R d_\Sigma\left(\left(\mu_{\tilde{P}}(x), \nu_{\tilde{P}}(x)\right), \left(\mu_{\tilde{Q}}(x), \nu_{\tilde{Q}}(x)\right)\right) dx \tag{1}$$

3 Representation of IFSs $\tilde{P} and \ \tilde{Q}$ Over the Interval [a_1, a_3] [20]

$$\mu_{\tilde{P}}(x) = \begin{cases} \left(\frac{x-a_1}{a_2-a_1}\right)^m & if a_1 \pounds x \pounds a_2 \\ \left(\frac{a_3-x}{a_3-a_2}\right)^m & if a_2 \pounds x \pounds a_3 \\ 0 & \text{otherwise} \end{cases} \quad \text{and} \quad \mu_{\tilde{Q}}(x) = \begin{cases} \left(\frac{x-a_1}{a_2-a_1}\right)^n & if a_1 \le x \le a_2 \\ \left(\frac{a_3-x}{a_3-a_2}\right)^n & if a_2 \le x \le a_3 \\ 0 & \text{otherwise} \end{cases}$$

$$\nu_{\tilde{P}}(x) = \begin{cases} \left(\frac{a_2-x}{a_2-a_1}\right)^m & if a_1 \pounds x \pounds a_2 \\ \left(\frac{x-a_2}{a_3-a_2}\right)^m & if a_2 \pounds x \pounds a_3 \\ 0 & \text{otherwise} \end{cases} \qquad \nu_{\tilde{Q}}(x) = \begin{cases} \left(\frac{a_2-x}{a_2-a_1}\right)^n & if a_1 \le x \le a_2 \\ \left(\frac{x-a_2}{a_3-a_2}\right)^n & if a_2 \le x \le a_3 \\ 0 & \text{otherwise} \end{cases} \tag{2}$$

Now as per Eq. (1), the pseudo-metric between two IFSs is given by

$$\begin{aligned} d^p\left(\tilde{P}, \tilde{Q}\right) &= \frac{1}{a_3 - a_1} \int_{a_1}^{a_3} d_\Sigma\left(\left(\mu_{\tilde{P}}(x), \nu_{\tilde{P}}(x)\right), \left(\mu_{\tilde{Q}}(x), \nu_{\tilde{Q}}(x)\right)\right)dx \\ &= \frac{1}{a_3 - a_1} \int_{a_1}^{a_2} d_\Sigma\left(\left(\mu_{\tilde{P}}(x), \nu_{\tilde{P}}(x)\right), \left(\mu_{\tilde{Q}}(x), \nu_{\tilde{Q}}(x)\right)\right)dx \\ &+ \frac{1}{a_3 - a_1} \int_{a_2}^{a_3} d_\Sigma\left(\left(\mu_{\tilde{P}}(x), \nu_{\tilde{P}}(x)\right), \left(\mu_{\tilde{Q}}(x), \nu_{\tilde{Q}}(x)\right)\right)dx \\ &= \frac{1}{a_3 - a_1} \int_{a_1}^{a_2} d_\Sigma\left(\left(\left(\frac{x - a_1}{a_2 - a_1}\right)^m, \left(\frac{a_2 - x}{a_2 - a_1}\right)^m\right), \left(\left(\frac{x - a_1}{a_2 - a_1}\right)^n, \left(\frac{a_2 - x}{a_2 - a_1}\right)^n\right)\right)dx \\ &\frac{1}{a_3 - a_1} \int_{a_2}^{a_3} d_\Sigma\left(\left(\left(\frac{a_3 - x}{a_3 - a_2}\right)^m, \left(\frac{x - a_2}{a_3 - a_2}\right)^m\right), \left(\left(\frac{a_3 - x}{a_3 - a_2}\right)^n, \left(\frac{x - a_2}{a_3 - a_2}\right)^n\right)\right)dx \\ &+ \frac{1}{a_3 - a_1} \\ &\int_{a_2}^{a_3} \max\left\{ \begin{array}{l} \frac{1}{2}\left|\left(\frac{x-a_2}{a_3-a_2}\right)^{2m} - \left(\frac{x-a_2}{a_3-a_2}\right)^{2n}\right|, \frac{1}{2}\left|\left(\frac{a_3-x}{a_3-a_2}\right)^{2m} - \left(\frac{a_3-x}{a_3-a_2}\right)^{2n}\right|, \\ \frac{1}{2}\left|\left(1 - \left(\frac{x-a_2}{a_3-a_2}\right)^m - \left(\frac{a_3-x}{a_3-a_2}\right)^m\right)^2 - \left(1 - \left(\frac{x-a_2}{a_3-a_2}\right)^n - \left(\frac{a_3-x}{a_3-a_2}\right)^n\right)^2\right| \end{array} \right\} dx \end{aligned} \tag{3}$$

Similarly, for the metric distance we have

$$\begin{aligned} d\left(\tilde{P}, \tilde{Q}\right) &= \frac{1}{a_3 - a_1} \int_{a_1}^{a_3} d_\Sigma\left(\left(\mu_{\tilde{P}}(x), \nu_{\tilde{P}}(x)\right), \left(\mu_{\tilde{Q}}(x), \nu_{\tilde{Q}}(x)\right)\right)dx \\ &= \frac{1}{a_3 - a_1} \int_{a_1}^{a_2} d_\Sigma\left(\left(\mu_{\tilde{P}}(x), \nu_{\tilde{P}}(x)\right), \left(\mu_{\tilde{Q}}(x), \nu_{\tilde{Q}}(x)\right)\right)dx \\ &+ \frac{1}{a_3 - a_1} \int_{a_2}^{a_3} d_\Sigma\left(\left(\mu_{\tilde{P}}(x), \nu_{\tilde{P}}(x)\right), \left(\mu_{\tilde{Q}}(x), \nu_{\tilde{Q}}(x)\right)\right)dx \\ &= \frac{1}{a_3 - a_1} \int_{a_1}^{a_2} d_\Sigma\left(\left(\left(\frac{x - a_1}{a_2 - a_1}\right)^m, \left(\frac{a_2 - x}{a_2 - a_1}\right)^m\right), \left(\left(\frac{x - a_1}{a_2 - a_1}\right)^n, \left(\frac{a_2 - x}{a_2 - a_1}\right)^n\right)\right)dx \end{aligned}$$

$$+\frac{1}{a_3-a_1}\int_{a_2}^{a_3} d_\Sigma\left(\left(\left(\frac{a_3-x}{a_3-a_2}\right)^m,\left(\frac{x-a_2}{a_3-a_2}\right)^m\right),\left(\left(\frac{a_3-x}{a_3-a_2}\right)^n,\left(\frac{x-a_2}{a_3-a_2}\right)^n\right)\right)\mathrm{d}x$$

$$=\frac{1}{a_3-a_1}\int_{a_1}^{a_2}\max\left\{\begin{array}{l}\left|\left(\frac{x-a_1}{a_2-a_1}\right)^m-\left(\frac{x-a_1}{a_2-a_1}\right)^n\right|,\\ \left|\left(\frac{a_2-x}{a_2-a_1}\right)^m-\left(\frac{a_2-x}{a_2-a_1}\right)^n\right|,\\ \left|\left(1-\left(\frac{x-a_1}{a_2-a_1}\right)^m-\left(\frac{a_2-x}{a_2-a_1}\right)^m\right)-\left(1-\left(\frac{x-a_1}{a_2-a_1}\right)^n-\left(\frac{a_2-x}{a_2-a_1}\right)^n\right)\right|\end{array}\right\}\mathrm{d}x$$

$$=\frac{1}{a_3-a_1}\int_{a_1}^{a_2}\max\left\{\begin{array}{l}\left|\left(\frac{x-a_1}{a_2-a_1}\right)^m-\left(\frac{x-a_1}{a_2-a_1}\right)^n\right|,\\ \left|\left(\frac{a_2-x}{a_2-a_1}\right)^m-\left(\frac{a_2-x}{a_2-a_1}\right)^n\right|,\\ \left|\left(1-\left(\frac{x-a_1}{a_2-a_1}\right)^m-\left(\frac{a_2-x}{a_2-a_1}\right)^m\right)-\left(1-\left(\frac{x-a_1}{a_2-a_1}\right)^n-\left(\frac{a_2-x}{a_2-a_1}\right)^n\right)\right|\end{array}\right\}\mathrm{d}x$$

$$+\frac{1}{a_3-a_1}\int_{a_2}^{a_3}\max\left\{\begin{array}{l}\left|\left(\frac{x-a_2}{a_3-a_2}\right)^m-\left(\frac{x-a_2}{a_3-a_2}\right)^n\right|,\\ \left|\left(\frac{a_3-x}{a_3-a_2}\right)^m-\left(\frac{a_3-x}{a_3-a_2}\right)^n\right|,\\ \left|\left(1-\left(\frac{x-a_2}{a_3-a_2}\right)^m-\left(\frac{a_3-x}{a_3-a_2}\right)^m\right)-\left(1-\left(\frac{x-a_2}{a_3-a_2}\right)^n-\left(\frac{a_3-x}{a_3-a_2}\right)^n\right)\right|\end{array}\right\}\mathrm{d}x$$

$$=\frac{1}{a_3-a_1}\int_{a_1}^{a_2} d_\Sigma\left(\left(\mu_{\tilde{P}}(x),\nu_{\tilde{P}}(x)\right),\left(\mu_{\tilde{Q}}(x),\nu_{\tilde{Q}}(x)\right)\right)\mathrm{d}x$$

$$+\frac{1}{a_3-a_1}\int_{a_2}^{a_3} d_\Sigma\left(\left(\mu_{\tilde{P}}(x),\nu_{\tilde{P}}(x)\right),\left(\mu_{\tilde{Q}}(x),\nu_{\tilde{Q}}(x)\right)\right)\mathrm{d}x \tag{4}$$

3.1 Numerical Example

Let us consider the interval valued fuzzy set like $[a_1, a_3]$=[10, 20], and we consider two IFSs $\tilde{P} and \tilde{Q}$ developed by using (2) where $\langle a_1, a_2, a_3\rangle = \langle 10, 15, 20\rangle$.. Then utilizing (4) and (5), we have the following result stated in Table 1.

Table 1 shows that for each criteria, the pseudo-metric distances between two IFSs as (pseudo) metric space is always less than the distances with respect to the usual metric. Also, it is seen that criteria *B* and criteria *C* gives exactly same results. Naturally, it comes because of the symmetric property of pseudo-metric spaces and metric spaces also.

Table 2 shows the ranking of distances over IFSs consisting of m–n exponts fuzzy ss under various criteria (integer-fraction). The ranking is getting ascending ord when the exponents are both integer (fraction) and keeps descending order when exponents of two different IFSs assume integer-fraction values. But in each case ranking type is same for pseudo-metric distance and metric distance. The distance under pseudo-metric spaces gives ranking of each distances over several criteria extensively, whereas the metric distance gives a broader sense of ranking. This means that for any kind of decision-making problems, if we want to measure a qualitative

Table 1 (Pseudo) metric distances under IFSs

Criteria	Exponents			Pseudo-metric distance	Metric distance
		m	n		
A	A_1	2	4	0.146	0.267
(m, n integer)	A_2	3	5	0.108	0.167
	A_3	4	6	0.079	0.114
B	B_1	2	1/2	0.150	0.7
(m integer n fraction)	B_2	3	1/3	0.228	1.000
	B_3	4	1/4	0.278	1.200
C	C_1	1/2	2	0.150	0.667
(m fraction n integer)	C_2	1/3	3	0.228	1.000
	C_3	1/4	4	0.278	1.200
D	D_1	1/2	1/4	0.126	0.267
(m, n fraction)	D_2	1/3	1/5	0.096	0.167
	D_3	1/4	1/6	0.074	0.114

difference between two subjects under study the pseudo-metric distance is more popular (user friendly) and easy to interpret for a decision-maker instead of metric distance only.

Figure 1 shows that the distances of two IFSs under metric space is higher than the distances under pseudo-metric when both the exponents of fuzzy numbers are integers. The minimum distance gap ranges from 0.04 to 0.14 approximately whenever we consider the exponents of fuzzy numbers as incremental indices

Figure 2a shows that the distances of two IFSs under metric is higher than the distances under pseudo-metric when one of the exponents of fuzzy numbers is integer and another one is fraction number.

Figure 2b shows the symmetricity of distances of (pseudo) metric spaces whenever the values of exponents of fuzzy numbers are getting interchanged.

Figure 3 expresses the distances of two IFSs under metric is higher than the distances under pseudo-metric when both the exponents are considered as fraction numbers

Figure 4shows that the variations of distances in various criteria of exponents, i.e. integer, fraction, both, etc., under (pseudo) metric. Also, it is clear from the graph that the distances of two IFSs as (pseudo) metric space are not intersecting. The gaps are getting increased when one index is integer and another is fraction than both indices are integer or fractions exclusively

Table 2 Ranking of (pseudo)metric distances

Criteria	Pseudo-metric distance	Metric distance
m, n integer	$A_1 > A_2 > A_3$	$A_1 > A_2 > A_3$
m integer n fraction or m fraction n integer	$B_1 < B_2 < B_3$ $C_1 < C_2 < C_3$	$B_1 < B_2 < B_3$ $C_1 < C_2 < C_3$
m, n fraction	$D_1 > D_2 > D_3$	$D_1 > D_2 > D_3$
Pseudo-metric distance	$B_3 > B_2 > B_1 > A_1 > D_1 > A_2 > D_2 > A_3 > D_3$	
Metric distance	$B_3 > B_2 > B_1 > A_1 > A_2 > A_3$	

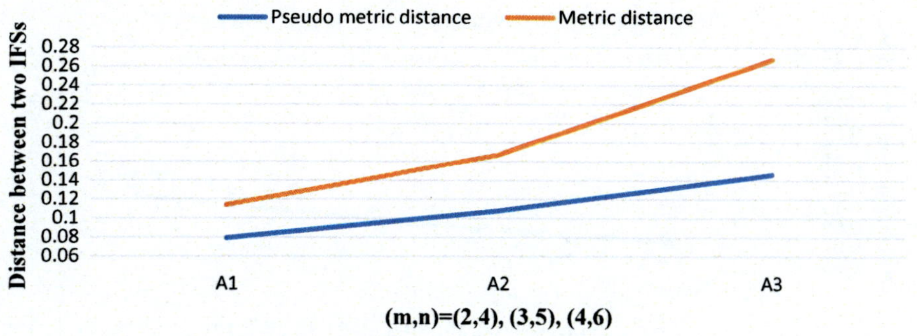

Fig. 1 Comparative study of distances when both exponents are integer

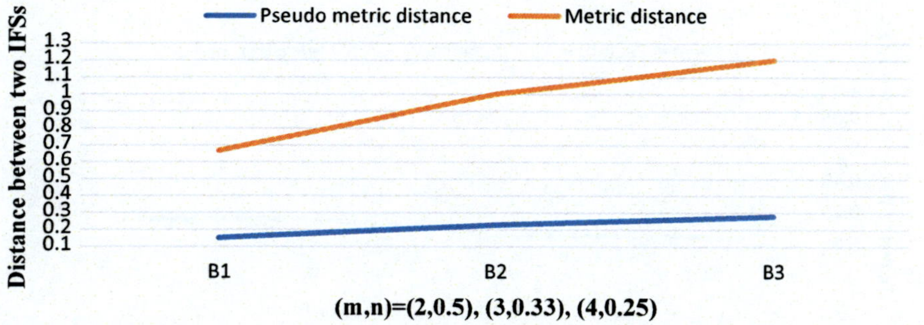

Fig. 2 **a**Comparative study of distances when both exponents is integer and anthor is fraction, **b** Comparative study of distances when both exponents is fraction and anthor is integer

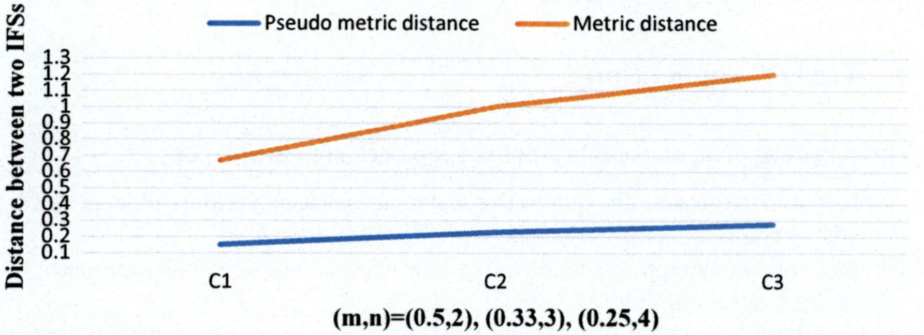

Fig. 2 (continued)

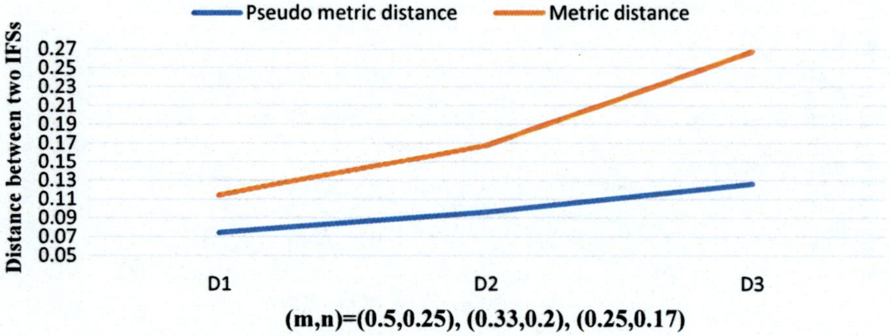

Fig. 3 Comparative study of distances when both exponents are fraction

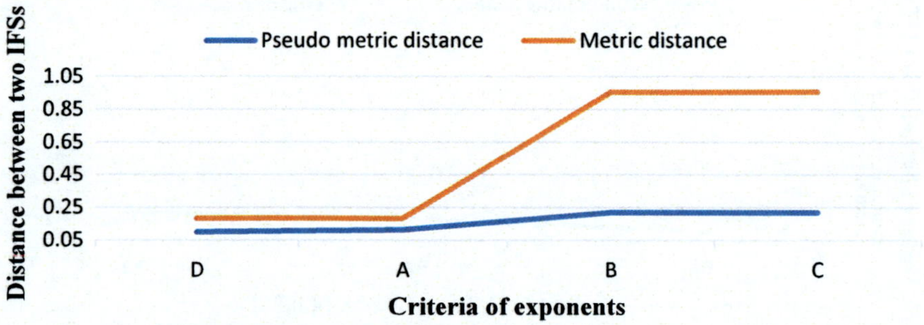

Fig. 4 Comparative study of distances for each criteria of exponents

3.2 Graphical Illustration

4 Area of Application

The area of applications of this proposed approach are stated below:

i. It is used to measure the qualitative differences among various subjects like any supply chain modelling or decision-making problem.
ii. To know the degrees of membership, non-membership and indeterminacy, this method can be applied.
iii. Any kind of ranking of different subjects over several disciplines is possible with its help.

4.1 Merits and Demerits

After the study of numerical and graphical illustrations, we see that there exists some merits and demerits of the proposed approach. They are stated as follows:

4.1.1 Merits

i. This approach is very useful to find difference between two IFSs whose membership, non-membership and indeterminacy functions are nonlinear.
ii. This method told us that the pseudo-metric distance is more user friendly for decision-making.
iii. This method helps us to study different nature of intuitionistic fuzzy sets drawn over physical problems with more detailing.

4.1.2 Demerits

i. This method is not applicable for the IFSs which are defined in a discrete space.
ii. This method is silent for higher dimensional intuitionistic fuzzy sets.
iii. This method might be complicated to handle when the membership function is complicated.
iv. The results may vary when the IFSs are assumed to be different.

5 Conclusion

In this study, we have discussed about the qualitative differences of the subjects of real-world problem by means of metric distances between two nonlinear intuitionistic fuzzy sets. Here, we see that the distance under pseudo-metric is more effective and extensive rather than the distance under conventional standard metric. The ranking of (pseudo) metric distances gives us a clear idea of quality measurement of two IFSs with nonlinear membership and non-membership function. This idea will be very useful to solve a decision-making problem. Also, graphical illustrations show the fluctuation of differences in several criteria of exponents in membership and non-membership function.

Scope of future work

This study of IFS with the help of metric and pseudo-metric is innovative. In future, various types of works can be done using this approach. This method can be applied for decision-making problems such as supply chain modelling or inventory modelling.

6 Conflicts of Interest

It is declared by the authors that there is no conflict of interest regarding the publication of this article.

References

1. Zadeh, L.A.: Fuzzy sets. Inf. Control **8**(3), 338–356 (1965)
2. Bellman, R.E., Zadeh, L.A.: Decision making in a fuzzy environment. Manag. Sci. **17**(4), 141–164 (1970)
3. Piegat, A.: A new definition of fuzzy set. Appl. Math. Comput. Sci. **15**(1), 125–140 (2005)
4. De, S.K., Beg, I.: Triangular dense fuzzy neutrosophic sets. Neutrosophic Sets Syst. **13**, 1–12 (2016)
5. De, S.K., Beg, I.: Triangular dense fuzzy sets and new defuzzication methods. J. Intell. Fuzzy Syst. (2016). https://doi.org/10.3233/IFS-162160
6. De, S.K.: Triangular dense fuzzy lock set. Soft. Comput. (2017). https://doi.org/10.1007/s00500-017-2726-0
7. Maity, S., Chakraborty, A., De, S.K., Mondal, S.P., Alam, S.: A comprehensive study of a backlogging EOQ model with nonlinear heptagonal dense fuzzy environment. RAIRO-Oper. Res. (2018). https://doi.org/10.1051/ro/201811
8. Maity, S., De, S.K., Mondal, S.P.: A study of an EOQ model under Lock Fuzzy Environment. Mathematics. (2019). https://doi.org/10.3390/math7010075
9. De, S.K., Mahata, G.C.: A comprehensive study of an economic order quantity model under fuzzy monsoon demand. Sadhana (2019). https://doi.org/10.1007/s12046-019-1059-3
10. Baez-Sancheza, A.D., Morettib, A.C., Rojas-Medarc, M.A.: On polygonal fuzzy sets and numbers. Fuzzy Sets Syst. **209**, 54–65 (2012)
11. Ban, A.I., Coroianu, L.: Existence, uniqueness and continuity of trapezoidal approximations of fuzzy numbers under a general condition. Fuzzy Sets Syst. **257**, 3–22 (2014)
12. Chutia, R., Mahanta, S., Baruah, H.K.: An alternative method of finding the membership of a fuzzy number. Int. J. Latest Trends Comput. **1**, 69–72 (2010)
13. De, S.K., Mahata, G.C.: Decision of a fuzzy inventory with fuzzy backorder model under cloudy fuzzy demand rate. Int. J. Appl. Comput. Math. (2016). https://doi.org/10.1007/s40819-016-0258-4
14. De, S.K., Pal, M.: An intelligent decision for a bi-objective inventory problem. Int. J. Syst. Sci. Oper. Logists. (2015). https://doi.org/10.1080/23302674.2015.1043363
15. Mahanta, S., Chutia, R., Baruah, H.K.: Fuzzy arithmetic without using the method of a-cuts. Int. J. Latest Trends Comput. **1**, 73–80 (2010)
16. Mao, X.B., Hu, S.S., Dong, J.Y., Wan, S.P., Xu, G.L.: Multiattribute group decision making based on cloud aggregation operators under interval valued hesitant fuzzy linguistic environment. Int. J. Fuzzy Syst. (2018). https://doi.org/10.1007/s40815-018-0495-2
17. Atanassov, K.T.: Intuitionistic fuzzy sets. VII ITKR's Session, Sofia (1983)
18. Atanassov, K.T.: Intuitionistic fuzzy sets. Fuzzy Sets Syst. **20**, 87–96 (1986)
19. Das, P., De, S.K., Sana, S.S.: An EOQ model for time dependent backlogging over idle time: a step order fuzzy approach. Int. J. Appl. Comput. Math. (2014). https://doi.org/10.1007/s40819-014-0001-y
20. Maity, S., De, S.K., Mondal, S.P.: A study of a backorder EOQ model for cloud-type intuitionistic dense fuzzy demand rate. Int. J. Fuzzy Syst. (2019). https://doi.org/10.1007/s40815-019-00756-1
21. De, S.K., Sana, S.S.: The (p, q, r, l) model for stochastic demand under intuitionistic fuzzy aggregation with Bonferroni mean. J. Intell. Manuf. (2016). https://doi.org/10.1007/s10845-016-1213-2
22. Deli, I., Broumi, S.: Neutrosophic soft matrices and NSM-decision making. J. Intell. Fuzzy Syst. **28**(5), 2233–2241 (2015)

23. Kaur, A., Kumar, A., Appadoo, S.S.: A note on '"approaches to interval intuitionistic trapezoidal fuzzy multiple attribute decision making with incomplete weight information."' Int. J. Fuzzy Syst. (2019). https://doi.org/10.1007/s40815-018-0581-5
24. Liang, R., Wang, J.Q.: A linguistic intuitionistic cloud decision support model with sentiment analysis for product selection in E-commerce. Int. J. Fuzzy Syst. (2019). https://doi.org/10.1007/s40815-019-00606-0
25. Xu, Z.S.: Intuitionistic fuzzy aggregation operators. Trans. Fuzzy System. **15**, 1179–1187 (2007)

Complex Structure of Number in Language Processing

Harjit Singh

Abstract This paper is particularly focused on plurality to discuss number system in Punjabi and Hindi. Taking the account of Punjabi, many interesting facts are found related to *'e'* marker. Scholars like Greville and Mithun (1996), (Rullmann (eds), Where Semantics Meets Pragmatics, Elsevier, Amsterdam, (2003), Kouider (eds.) (Kouider, S. (eds.).: Acquisition of English number marking: the singular and plural distinction. Language Learn Dev. 2(1), 1–25 (2006)), (Alexiadou, A. Plural. In: Proceedings of the 28th West Coast Conference on Formal Linguistics) and (Scontras, G.: A: In: Chemla, E., Homer, V., Winterstein, G. eds.) Proceedings of sub 17, pp. 545–562) have widely demonstrated the nature and use of associative plurals, indefinites; mass and count nouns and numerals in the context of semantics, respectively. (Schwertel, U.: Plural semantics for natural language understanding- a computational proof theoretic approach, Ph.D. thesis, University of Zurich (2005)) has developed the disambiguate system within numbers, while Farkas and de Swart (Farkas, D. (eds.): the semantics and pragmatics of plurals. Semantics and Pragmatics 03 (2010)) identified the pragmatic factors to determine inclusive and exclusive plural forms in English. Schütze (1989) has found the important place of co-ordinators such as *'and'* in memory retrieving tasks. As usual numbers in Punjabi, which means talk about singular and plural in the case of kʊʈ̪a (dog) = singular and kʊʈ̪e(dogs) = plural. To deal plurality here, it is significant to focus on attributes such as *'e'* in some cases ((mõd̪e *'boy' = common noun,* e *'you' = pronoun,* ɟiʈ̪e *'name of a person' = noun,* vire *'brother'*, òne *'he'/ 'she' = pronoun,* səbe *'all' = quantifier* (∃), kale *'name of a person' = noun,* ha:le *'now' = adverb,* kʰèd̪e *'play' = verb,* pɔ̃ɟe *'five' = numerals,* nɪle *'blue' = adjective,* de-d̪e *'give' = postposition reduplication,* əte *'and' = coordination,* etc.) of Punjabi. Investigating such cases, it is argued that a marker *'e'* can be (singular, plural-cum-singular and plural) also to denote a person or an object. Further, it is possible to filter out the marker under some mapping conditions that followed by few notations (– / +). The proposal is to address that a marker *'e'* can be picked up

---

H. Singh (✉)
IGNTU, Amarkantak, M.P., India
e-mail: harjitsingh.jnu@gmail.com

O. Castillo et al. (eds.), *Applied Mathematics and Computational Intelligence*, Springer Proceedings in Mathematics & Statistics 413,
https://doi.org/10.1007/978-981-19-8194-4_5

within verbal or written discourse to match with an object (*O*). And further, an object is executed to carry some properties (*P*) to check singular (Sn) and plural (Pl) to control contextual knowledge (CK) in languages like Punjabi.

Keywords Number · Plural · Attributes · List · Variable · Mapping

1 Introduction

Greville and Mithun [9] have suggested the Smith-Stark hierarchy of associative plurals in Central Pomo and Central Alaskan Yup'ik to keep them into a separate category [17] have been generalised the Mandarin Chinese number system followed by bare nouns and indefinites. They found semantic-pragmatic properties among these nouns. Kouider (eds.) [11] has been working into design a new methodology for plural processing among the children. It was discovered that a framing trail got succeed to show the differences between singular and plural for 20 years old [2] has adopted a morph-syntax framework to define mass and count nouns [10] has found additive features in approximate and non-approximate numbers to see compositional morphology and semantics [22] has shown the semantic context of numerals in Turkish, Hungarian and Western Armenian [1] has explained that the number system could interface with syntax, morph-syntax and semantics.

1.1 Number Processing: A Background

In [20] has developed the disambiguate system of plurals followed by two Schütze [21] and Aone [3]. For discussing German quantifiers, Schütze [21] believes in Pafel's scale of distributivity. While on the other hand, Aone [3] has proposed the plural disambiguation algorithm based on external knowledge factors. Farkas and de Swart [7] made observations regarding the pragmatic factors to locate the inclusive and exclusive plural forms in English. Maldonado (eds.) [12] has conducted a study on some number of participants to explore plural ambiguities in terms of cumulative and distributives. Yatsushrio (eds.) [25] has proposed that plurals are morphologically marked but semantically unmarked in German. Schütze [21] has found the important place of co-ordinators such as *'and'* in memory retrieving tasks.

1.2 Previous Works

In this section, it is going to discuss about computational aspects of number cross-linguistically.

Table 1 Chickasaw nouns

Chickasaw nouns	Singular	Plural
Ihoo	Woman	Women
Ofi	Dog	Dogs
foni	Bone	Bones

1.2.1 Numbers in Computational Mechanism

It is not surprising to deal with different forms of number in different languages. Sometimes, existing structural variations among number may be triggered either similar or unrelated implementation conditions.

- ***Chickasaw Quantifiers***
  Munro ([14]: 125) has pointed out that following noun categories in Chickasaw do not have visible number marking. Below examples either can be a singular or as plural (Table 1)

Above set of examples in Chickasaw nouns indicate that the form of (*'ihoo, ofi and foni')* do not change into singular and plurals. To demonstrate the collective form of plurality, it is seen that the part as *'alhiha',* which means the 'bunch' can produce bunch meanings. For example,

(a) Ihoo-at taloowa
Woman-nom sing
'The woman sings'
(b) alhih-aat taloowa
Woman bunch-nom sing
'A bunch of women sing'

It is shown that *'alhiha'* produces the collection or bunch meanings in above two examples. In second (b), when *alhiha* comes between *ihoo* and *taloowa* to construct plurality here (as cited in Paperno and Keenan [15], p. 125).

- ***Arabic Broken Plural Generation System-***
  Blanchete et al. [4] have presented the generation system for broken plurals (BPs) in Arabic. They found that linguistic features (like phonological, morphological and semantic) around 108 in singular forms (SF) using to control the entire process. On the other hand, there is also an important place of root-pattern methodology here (as cited in Ignazio Mauro Mirto (eds.) [13]: p 27).

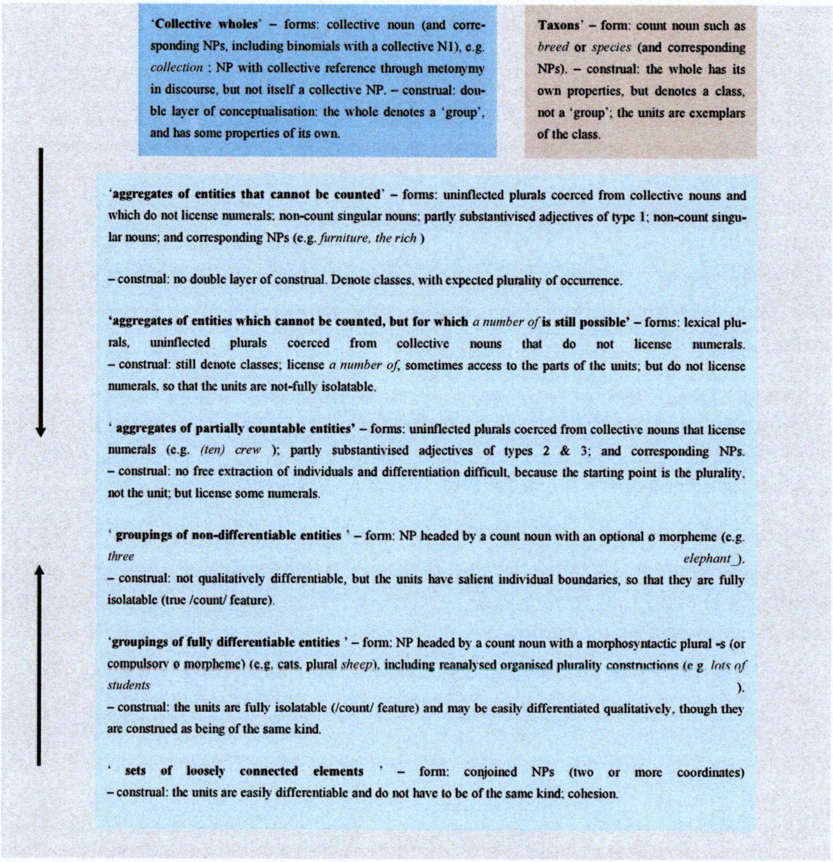

Fig. 1 Collective wholes and Taxons

- ***Scale of Unit Integration for Pluralities of Entities-***

  Gardelle [8: 186–89] has defined the utility of plurals in terms of designing the unit integration scale in English. The following steps in a figure can be generalised as-

Figure 1, it is indicated that the use of 'collective whole' means to focus on basically collective nouns and 'taxons' introducing the quantity of nouns. Some cases like (furniture, the rich) are uncountable plurals that cannot be inflected. In these cases, it may be trying to extract numerals in uninflected forms by applying them at the beginning, as in 'ten crew', etc. While in groups, Ø morpheme can be voluntary but it morph-syntactically converts as a plural marker -s, in Table 2.

Table 2 Singular-plural marking

ਆ (Singular)	ਏ (Plural)	ਈ (Singular)	ਆਂ (Plural)
ਕੁੱਤਾ(Dog) kʊt̪a	ਕੁੱਤੇ(Dogs) kʊt̪e	ਕੁੜੀ(Girl) kʊɽi	ਕੁੜੀਆਂ (Girls) kʊɽiɑ̃
ਘੋੜਾ(Horse) kòɽa	ਘੋੜੇ(Horses) kòɽe	ਆਦਮੀ (Man) ɑːd̪əmi	ਆਦਮੀਆਂ (Men) ɑːd̪əmĩ
ਮੁੰਡਾ(Boy) mʊ̃ɖa	ਮੁੰਡੇ(Boys) mʊ̃ɖe	ਕੁਰਸੀ (Chair) kʊrəsi	ਕੁਰਸੀਆਂ (Chairs) kʊrəsĩ
ਝੋਲਾ (Bag) cə́ola	ਝੋਲੇ(Bags) cə́ole		

2 Number System in Languages

There are two languages (Punjabi and Hindi) have been chosen for describing the number system. First, it would focus on Punjabi number system and secondly will investigate Hindi number system.

2.1 Punjabi Number System

Punjabi has two kinds of number representations such as (singular and plural) in general.[1] Punjabi is a modern Indo-Aryan language, which is less similar with Hindi. Both cannot be alike at different linguistic levels. There are two forms of gender male and female. Similarly, there are two forms of number (singular and plural) also in Punjabi.[2] It is 'ਆ' marked a singular and ਏ' to the plurals in Table 2.

Table 2 shows the number marking in Punjabi. ਆ' is singular indicator can be equal to singular forms like dog, horse, boy and bag in English. Plural information is given by ਏ' to signify + s in pairs like dog + s, horse + s, boy + s and bag + s, etc. It is significant to know that ਆ' and 'ਏ' are conventional signs in Punjabi number system.[3]

[1] For more details see pan-ind-styleguide.pdf.

[2] Vishawbharti@tdil.

[3] Http://www.learnpunjabi.org/noun.html

Table 3 Hindi plurals

आ (Singular)	ए (Plural)	अ (Singular)	एं (Plural)	ई (Singular)	इया (Plural)	या (Singular)	याँ (Plural)
घोड़ा (Horse)	घोड़े(Horses)	आँख (Eye)	आँखे (Eyes)	नारी (Lady)	नारियाँ (Ladies)	चिड़िया (Sparrow)	चिड़ियाँ (Sparrows)
कमरा (Room)	कमरे(Rooms)	पुस्तक (Book)	पुस्तकें (Books)	चाबी (Key)	चाबियाँ (Keys)	बुढ़िया (old lady)	बुढ़ियाँ (old ladies)
बेटा (Boy)	बेटे(Boys)	बहन (Sister)	बहनें (Sisters)	नदी (River)	नदियाँ (Rivers)	पुड़िया (Parcel)	पुड़ियाँ (Parcels)
रास्ता (Way)	रास्ते(Ways)						
कुत्ता (Dog)	कुत्ते(Dogs)						

2.2 *Hindi Number System*

Like Punjabi, there are two basic forms of number system can be found in Hindi. 'आ' and 'ई' both are common signs to indicate singular and plural forms (Table 3).[4]

In this Table 3, there are four parallel sets of singular-plural forms in Hindi. 'आ' marks the singular and'ए' indicates the plural in the first set. Second set'अ' identifies a single form of आँखआँखे,पुस्तक and बहन, whereas 'एं'' carries the plural form of नदी, पुस्तकें and बहनें in Hindi. 'ई' is a singular form in नारी, चाबी and नदी but 'इया' gives plural form like नारियाँ, चाबियाँ and नदियाँ in the third set. Fourth set introduces the singular form with 'या' in चिड़िया, बुढ़िया and पुड़िया similarly the plural form comes with 'याँ' in, चिड़ियाँ, बुढ़ियाँ and पुड़ियाँ.

3 Plural Structure: Puzzles and Attributes

Russell and Whitehead [19] said that the definite descriptions of 'the' can be interchanged with classes. In monadic first order logic, the plurals identified by Bools. Rumfitt [18] has also drawn a similar investigation in English. Number can be translated into singular and plural with variables representations like (*x, y, z* and **x, y, z**) (Table 4).[5]

---

[4] https://blogs.transparent.com/hindi/how-to-make-plural-from-singular-noun-in-hindi/

[5] See Smiley and Oliver [24] pp. 106–10.

Table 4 Plural variables

Plural variables	Valuation of Variables	Plural quantifiers
Simons [23] ***x y z***	Any number of things	∃, ∀
h k l	Two or more things	∃, ∀
Burgess and Rosen [5] *xx yy zz*	One or more things	∃∃, ∀∀∃, ∀
Rayo [16] *xx yy zz*	One or more things	∃, ∀
Yi [26] *xs ys zs*	One or more things	Σ, Π

Variables *x*, *y*, *z* can stand for persons, things, etc., in a discourse. Quantifiers, such as ∃ (existential) and ∀ (all) can be used for plurality in natural language.

3.1 Attributes About ‘e’?

It is difficult to understand the nature of (kʊʈe *’e’*) in Punjabi. Because sometimes, it functions like an attribute and sometimes does not. However, this is not the case of Hindi. In general, the tendency of *‘e’* denoting plurals when it comes at the end of nominal categories. In some cases, it may give different results without changing the form of *‘e’*. In Punjabi, there are number of variants can be possible with *‘e’*. Below see the list (Table 5).

I. **mʊ̃ɖe** *‘boy’ = common noun.*
II. **e** *‘you’ = pronoun.*
III. **ɟiʈe** *‘name of a person’ = noun.*
IV. **vire** *‘brother’.*
V. **òne** *‘he’/ ‘she’ = pronoun.*
VI. **sə̀be** *‘all’ = quantifier (∃).*
VII. **kale** *‘name of a person’ = noun.*
VIII. **ha:le** *‘now’ = adverb.*
IX. **kʰèɖe** *‘play’ = verb.*
X. **pə̃ɟe** *‘five’ = numerals.*
XI. **nɪle** *‘blue’ = adjective.*
XII. **ɖe-ɖe** *‘give’ = postposition reduplication.*
XIII. **əʈe** *‘and’ = coordination.*

Table 5 Linguistic description of 'ੲੇ' '*e*''

IPA Transcription	**English Interpretation**	**Description of '*e*'**
1. **mõḍe** nõ bʊlao	'Call the boy'	**mõḍe** *'Boy'= Common Noun*
2. **e̱**a: ɟa:	'You come here'	**e** *'You'= Pronoun*
3. **ɟite** kiṯʰe gəya si	'Jeete, where did you go?'	**ɟite** *'Name of a person'=Noun*
4. **vire** a: ɟa:	'Come here, Brother'	**vire** *'Brother'*
5. **òne** kʊ́ɟə nəhi keha	'He/She did not say anything'	**òne** *'he'/'she'= Pronoun*
6. **sə̀be** a: ɟao	'All come here'	**sə̀be** *'All'= Quantifier (∃)*
7. **kale** ḏe piṯa ɟi kɪ kərḏe ne	'What is doing Kale's father?'	**kale** *'name of a person'= Noun*
8. **ha:le** rʊk ɟao	'Now stop here'	**ha:le** *'Now'= Adverb*
9 oh **kʰèḏe** bɪna ʃəla gəya	'He went back without a play'	**kʰèḏe** *'Play'= Verb*
IPA Transcription	**English Interpretation**	**Description of '*e*'**
10. **pə̃ɟe** gəllã: səʃĩ: ne	'Five things are correct'	**pə̃ɟe** *'Five'= Numerals*
11. **de-de** pɑ́:i	'Give me brother'	**de-de** *'Give'= Postposition Reduplication*
12.rɑ:m **əte** ʃɑ:m pá:i ne	'Ram and Sham are brothers'	**əte** *'and'= Co-ordinator*

4 Proposal for Plural Processing

Using the variants of *'e'* in above list seems more productive in terms of different word classes. Notice that the semantics played an important role from nominal to co-ordinators in natural language.[6] It is an interesting to study the category patterns of 'e' in Punjabi. Look at Table 6.

It is a clearly shown that the occurrence of *'e'* in all examples can assume as a 'list' here. It is productive as mentioned above. Variables (*x*, *y*) have been given to each list. Again, variables tagged with (*Xs* and *Ys*) symbols to process the plurals.

[6] Davidson [6] 'Semantics of Natural Language'.

Table 6 List, productive, explanations, and symbols

List	Productive	Explanations	Symbol
e'	*x y*	Variables (*S* and *P*)	Xs Ys
e'	*x y*	Variables (*S* and *P*)	*Xs* Ys
e'	*x y*	Variables (*S* and *P*)	*Xs Ys*
e'	*x y*	Variables (*S* and *P*)	*Xs Ys*
e'	*x y*	Variables (*S* and *P*)	*Xs Ys*
e'	*x y*	Variables (*S* and *P*)	*Xs Ys*
e'	*x y*	Variables (*S* and *P*)	*Xs Ys*
e'	*x y*	Variables (*S* and *P*)	*Xs Ys*
e'	*x y*	Variables (*S* and *P*)	*Xs Ys*
e'	*x y*	Variables (*S* and *P*)	Xs Ys
e'	*x y*	Variables (*S* and *P*)	Xs Ys

5 Modelling Presupposition and Analysis

Followed by a model, it is said that an input gives a sign like *'e'*, which can be used as singular, plural-cum-singular and plural according to persons and things in the first stage. Next stage, it shows the integration with an object and a concept through mapping conditions.

In terms of attributes, different variants of '*e*' can select different arithmetic val-ues. In a similar way, each attribute can be matched with variables using *x* and *y* symbols. All this information can be inserted into a model to filter out the steps applied in first and second stages until reach the end.

6 Conclusion and Future Work

It is concluded that each object or an article takes number features either can be a singular or plural and may be something else. The case of *'e'* in Punjabi fulfils the condition that the plural can behave like a singular. Previously, it has been mentioned that *'e'* has many variants according to word classes. In plural processing, the list of *'e'* produces singular, plural-cum-singular, and plural forms for denoting objects around us. Plurality among the objects can be measured with using notations (±). It is simple to understand that a user who picks up an object can be marked with the *'e'* to suggest a singularity followed by the properties like (Sn/Pl) depending on the context. In short, the usage of '*e'* firstly considers as a plural in very common word pairs, as it has been discussed into Table 2. On the other hand, it can be generalised based on a Table 5 analysing the *'e'* linguistically. Next, it is tried to draw an idea under the Fig. 2, defining the model to capture the entire process from beginning to end. In future, by incorporating the large data, based on

Table 7 Attributes of *'e'*

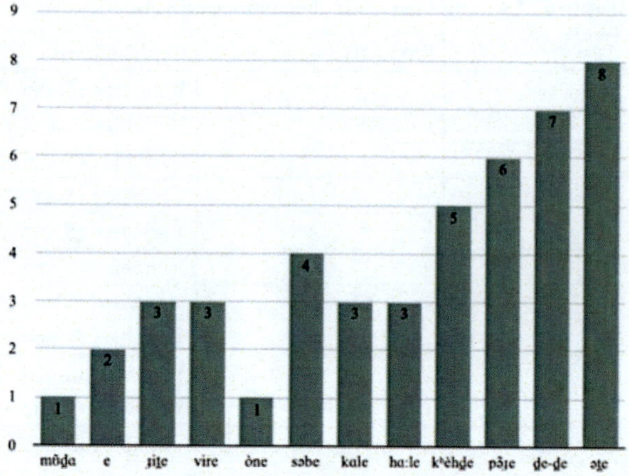

Table 8 List, variables and symbols

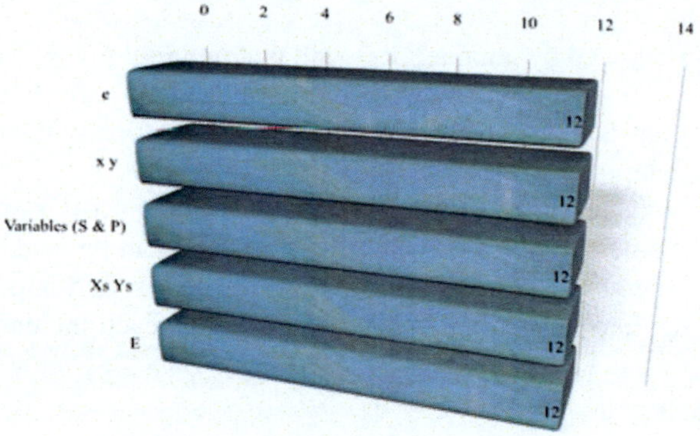

other markers like 'ਆਂ' to be used to find out the same generalisation. It would also be attempted to design an algorithm by surveying such data to further go for a system development especially in Punjabi.

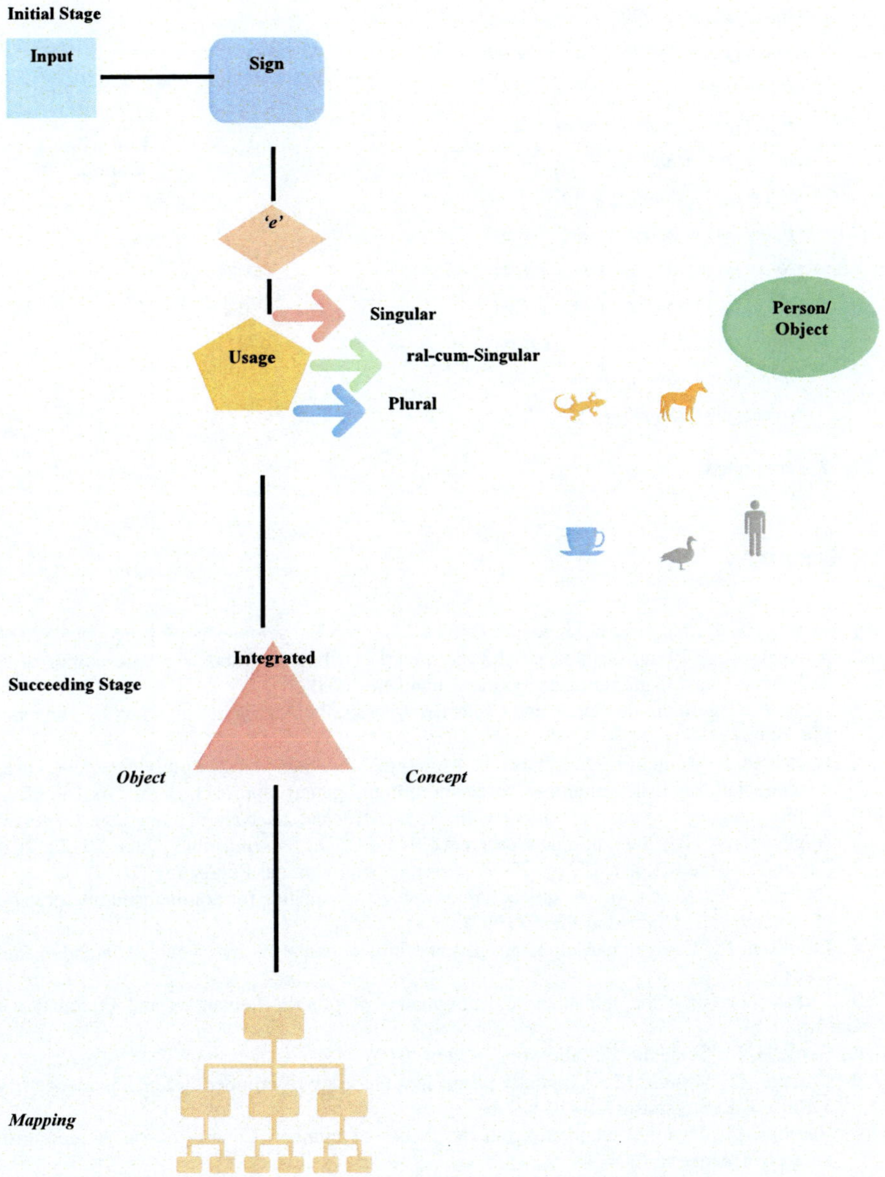

Fig. 2 Modelling number system

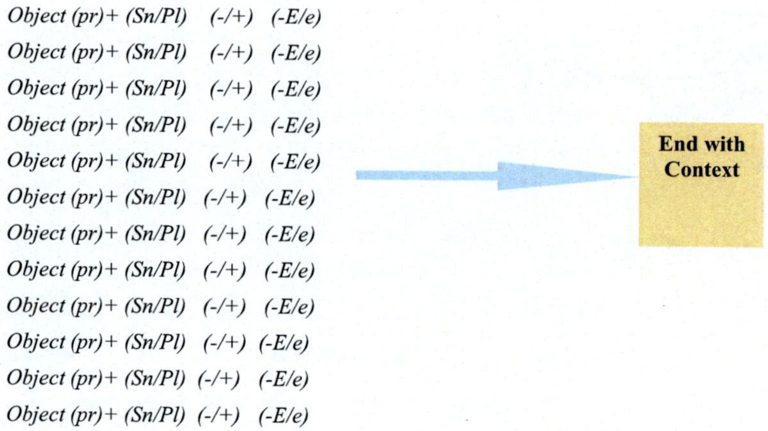

Fig. 2 (continued)

References

1. Acquaviva, P.: Number in language, Oxford Research Encyclopaedia of Linguistics (2017)
2. Alexiadou, A.: Plural, abundance and the theory of Murpho-syntax. In: Proceedings of the 28th West Coast Conference on Formal Linguistics (2011)
3. Aone, C.: Treatment of Plurals and Collective-distributive Ambiguity in Natural... University Microfilms (1991)
4. Blanchete, I., Mourchid, M., Mbarki, S., Mouloudi, A.: Arabic broken plural generation using the extracted linguistic conditions based on root and pattern approach in the NooJ Platform. In: Formalizing Natural Languages with NooJ 2018 and Its Natural Language Processing Applications: 12th International Conference, NooJ 2018, Palermo, Italy, June 20–22, 2018, Revised Selected Papers 12, pp. 27–37. Springer International Publishing (2019)
5. Burgess, J. P., Rosen, G.: A subject with no object: Strategies for nominalistic interpretation of mathematics. Clarendon Press (1997)
6. Davidson, D.: Essays on Actions and Events: Philosophical Essays, vol. 1. Clarendon Press (2001)
7. Farkas, D. (eds.): the semantics and pragmatics of plurals. Semantics and Pragmatics **03** (2010)
8. Gardelle, L.: Semantic Plurality, pp. 1–225 (2019)
9. Greville, C., Mithun, M.: Associative form in a typology of number systems: evidence from Yu'pik. J. Linguistics **32**, 1–17 (1996)
10. Harbour, D.: Paucity, abundance and the theory of number. Language vol. 90, Linguistic Society of America (2014)
11. Kouider, S. (eds.).: Acquisition of English number marking: the singular and plural distinction. Language Learn Dev. **2**(1), 1–25 (2006)
12. Maldonado, M. (eds.): Priming plural ambiguities, J. Memory Language **95**, 89–101 (2017)
13. Mauro, M. (eds.): In Formalizing Natural Languages with NooJ 2018 and Its Natural Language Processing Applications: 12thInternational Conference, NooJ 2018, Palermo, Italy, June 20–22, 2018, Revised Selected Papers 12, pp. 27–37. Springer International Publishing, 2019
14. Munro, P.: Chickasaw quantifiers. In: Handbook of Quantifiers in Natural Language, vol. II, pp. 113–201 (2017)

15. Paperno, D., Keenan, E. L. (eds.): Handbook of quantifiers in natural language: Volume II, vol. 97. Springer (2017)
16. Rayo, A.: Word and objects. Noûs **36**(3), 436–464 (2002)
17. Rullmann, H., You, A.: General number and the semantics and pragmatics of indefinite bare nouns in Mandarin Chinese, In Von Heusinger, K., Turner, K. (eds.) Where Semantics Meets Pragmatics, pp. 175-196 Amsterdam: Elsevier (2003)
18. Rumfitt, I.: Plural terms: another variety of reference?. Thought, reference, and experience: Themes from the philosophy of Gareth Evans, pp. 8–123 (2005)
19. Russell, B., Whitehead, A. N., Frege, G., Peano, G.: Principia mathematica defines the logistic movement. In, Great Events from History: The 20th century, 1901–1940, vol. 2, p. 814 (2007)
20. Schwertel, U.: Plural Semantics for Natural Language Understanding—A Computational Proof Theoretic Approach, Ph.D. thesis, University of Zurich (2005)
21. Schütze, H.: Pluralbehandlung in natürlichsprachlichen Wissensverarbeitungssystemen. Inst. Für Wissensbasierte Systeme (1989)
22. Scontras, G.: A unified semantics for number marking, numerals, and nominal structure. In: Proceedings of Sub 17, ed. Emmanuel Chemla, Vincent Homer and Gregorie Winterstein, pp. 545–562. Paris, France: Semanticsarchive (2013)
23. Simons, P.: Plural reference and set theory. Harvard (1982)
24. Smiley, T., Oliver, A.: Plural logic (2013)
25. Yatsushiro, K. (eds.).: The Unmarkedness of plural: cross-linguistic data. In: LaMendola, M., Scott, J.: Proceedings of the 41st Annual Boston University Conference on Language Development, pp. 753–765. Somerville, M.A: Cascadilla Press (2017)
26. Yi, B. -U.: The logic and meaning of plurals. Part II. J. Philos. Log. **35**, 239–288 (2006)

Digital Newspaper Using Augmented Reality

Ayrin George, Divya A. Pillai, Joel Joseph, and Smita Rukhande

Abstract Augmented reality (AR) is a recent technology that provides the users with an enhanced and interactive experience with the digital world, by combining real-time objects and computer-generated images and videos. This project aims to focus on the integration of AR with the newspaper. Since the daily newspapers provide only limited content about an article, this application will allow the users to have detailed information about an event with the help of video clips and interactive images. The application provides a feature where the users will be redirected to the online website as soon as they click on the media on their screen, where they can learn more about the respective commercial and the user can bookmark an article. The process of linking media files with the article's image will be implemented using Unity and Vuforia. The application will be developed using Android Studio.

Keywords Augmented reality · Unity · Vuforia · Android studio · Newspaper

1 Introduction

A newspaper is a set of printed papers that have daily or weekly updates of current happenings around the world to provide readers with up to date information on current affairs. These are printed daily or weekly by an organization. The count of newspaper readers has been reduced across the world due to the introduction of digital news such as e-news or news feeds. This can be termed as the impact of digitalism. The reading habit of newspaper among youngsters has reduced gradually. Smartphones along with Internet facilities have reduced the impact of newspapers on people. The online coverage of the news is far more than what a single newspaper holds. At this

Supported by organization Fr Conceicao Rodrigues Institute of Technology, Vashi.

A. George (✉) · D. A. Pillai · J. Joseph · S. Rukhande
Information Technology, Fr. Conceicao Rodrigues Institute of Technology, Vashi, Navi Mumbai 400703, Maharashtra, India
e-mail: ayrin.george@gmail.com

O. Castillo et al. (eds.), *Applied Mathematics and Computational Intelligence*, Springer Proceedings in Mathematics & Statistics 413,
https://doi.org/10.1007/978-981-19-8194-4_6

rate, the next generation might not be seeing a newspaper at all. Though e-readers have a wide user base, many people still prefer printed newspapers.

To increase the count of users of the printed newspaper, we need to make newspapers much more interesting. This project is aimed at digitizing traditional newspapers with the use of augmented reality. This will not only make newspapers more interesting but will add more value and information to it.

In today's world, augmented reality [1] is one of the popular and highly efficient technologies which provide the users with an enhanced and interactive experience with the digital world, by combining real-time objects with computer-generated images and videos. Publications and marketing are increasingly using augmented reality to increase the interactivity of the users [2]. Communication through multiple channels and integrating the old style with the existing material, i.e., by combining digital and print media [3], creating a competitive advantage and bringing in new revenue can be achieved by AR which can boost the advertising market. Transformation of traditional newspapers which is an offline media into digital can be done through AR publications. With AR, the user will be able to view both content and illustration in one screen and a camera [4]. The newspaper will gain a new cutting edge along with an increase in the number of readers after the implementation of augmented reality.

2 Literature Survey

The features of augmented reality science textbooks [5] review several pieces of literature concerning the features of augmented reality (AR)-based textbook that could be applied for science learning in schools to make the learning process effective and interesting. Usually, science is considered as a difficult subject by the students to understand. So, considering this, AR is used in textbooks to make science more interesting and fun. This technology offers the students with ultimate imaginary and makes them experience as well as understand the concept behind science through video, audio, and graphics. This helps students to overcome the fear of science and as well as understand the concepts. The AR technology used here is marker-based, in this, an AR code is used as a key to uniquely identify the element to find a related video, audio, or graphics from the database. Our proposed system uses marker-based technology which does not have an AR code but instead uses any content available on the page to identify objects from the database. Moreover, the above system is only limited to the students who wish to study science but our system can be used by all the users who read the newspaper.

Marker-based AR has an inherent drawback, i.e., requirement of markers which may be costly and limit the development of the application. The markerless technology which uses visual or depth information of the captured scene top estimates the camera pose. A markerless AR framework is designed that simultaneously considers mobility, accuracy computation complexity. It can achieve real-time 3D environment reconstruction by using mobile equipped with a depth camera and Inertial Measurement Units [6]. But markerless tracking requires high computational power.

AR is also combined with virtual reality (VR) to create Magic Book [7] to help users immerse themselves in a completely different world. Here, users can just scan or view the contents of the book with specialized equipment to experience the features provided by the book. VR users can see other VR users represented as life-sized virtual avatars, while AR users will see VR users as miniature avatars in the scene.

A user study trends in augmented reality and virtual reality research [8] paper describe the trends of how user studies have been incorporated into AR and VR. Also, the author presents implications on what needs to be taken into account when planning a user study in the field of AR and VR research. Both AR and VR research tend to incorporate the user study into their research to test the effectiveness or efficiency of their research. It was discovered that the rate of conducting user studies in the past three years of AR and VR research was less than 50%. The rate tells us that AR and VR research need to work with their users more in their research. Both AR and VR research are for human users and failure to understand users' needs and requirements of AR and VR research will result in user frustration or cause an unsatisfactory experience. Therefore, user-based evaluations, either formative or summative or both of them, need to be taken into account in the design and development of AR and VR research to produce more usable outcomes for their users.

In the future, eBooks are expected to replace the traditional book and references as studying material because of the advances in Information Technology (IT) [9].

A mobile application has been developed combining the use of traditional printed media and augmented reality to boost the popularity of the traditional Malaysian games [10]. The mobile application provides interactive features to allow the user to interact and manipulate the 3D objects to increase the efficiency of the game.

According to the research conducted, print media is in crisis due to the digital media. Print media is losing its hold on advertisements—its main source of income to digital media. According to the author [11], the crisis of print media can be avoided by creating a hybrid version that is using AR technology in the print media.

The evolution of the smartphone has also fueled the use of augmented reality in a phone [12]. The smartphone plays an important role in the application of augmented reality if studied closely AR can be implemented in the phone using the application as a media and various features of the phone can be used for the implementation to improve the experience of AR.

3 Application Description

Digital newspaper is an Android application that is intended to provide users with add on information apart from the one present on the printed newspaper. This includes additional videos/animation related to particular news, article, advisement, or the user can directly open a website from their phone just by scanning related articles. This is

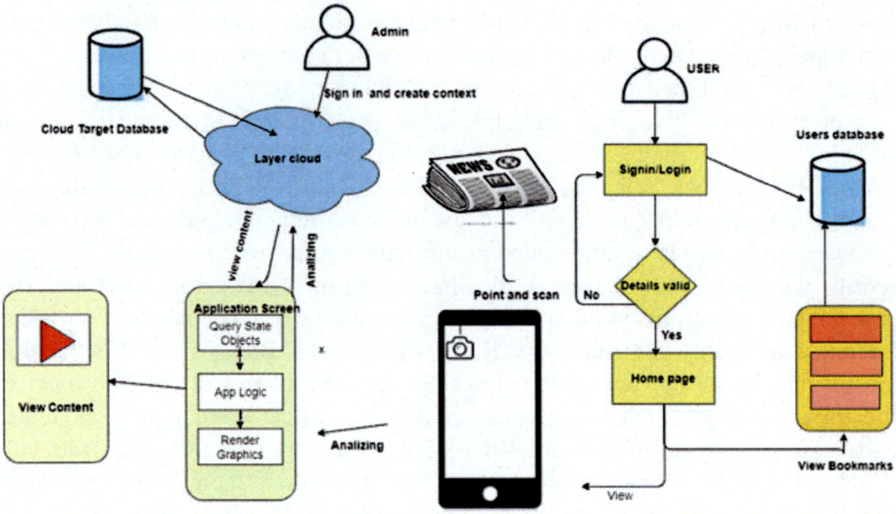

Fig. 1 Architectural diagram of digital newspaper

implemented using AR. The traditional newspaper lacks much of the information due to lack of space, etc. By scanning the contents of the newspaper, the users can view additional graphics, objects, videos, etc., on their mobile application. This is also used to make advertisements more interesting and engrossing. The digital newspaper also provides users with bookmarking facilities as shown in Fig. 1.

The main objective of digital newspapers to provide a hybrid version of the traditional print media and digital media, to reduce the declined number of users of the print media.

The components used for creating digital newspapers are smartphone or Android Phone, Android Studio, Unity3D, and Vuforia.

3.1 Smartphone or Android Phone

The smartphone is used by the user for launching and using the application which facilitates the digital newspaper functionality like scanning the 2D images present in the newspaper for advertising purposes or achieving additional news article related information. The camera of the smartphone is used for scanning, and related output can be seen on the smartphone screen.

3.2 Android Studio

Android Studio would be used for creating the Android application which would be run on the user's Android phone, through which the user can operate and use various functions.

3.3 Unity 3D

Unity 3D is a graphics and physics engine that is used to build scaleable applications that can be built for multiple platforms with the same codebase. Platforms supported by Unity include Linux-x86/x86-64, Mac-x86/x86-64, Windowsx86/x86-64, iOS, Android, and WebGL. Unity uses C for internal scripts and logic.

3.4 Vuforia

Vuforia is used to create an augmented reality application, and it can be considered as an augmented reality Software Development Kit for mobile devices. It provides detection and tracking of image targets by using feature detection, a feature is any point in an image. It was available as a plug-in for Unity. Vuforia has the following components

AR Camera: the camera can be used to track the objects.

Tracker Image: The tracker image is the most essential part of this SDK. A tracker image is required for a device's camera to recognize a reference and track it. The orientation and actual size of the tracker image directly affect the same attributes of the superimposed images. Any image can be assigned as a tracker image. The features of the target image effectively determine how well the target is tracked by the tracker. The JPEG or PNG images in RGB or grayscale are supported by image targets.

Markers: Vuforia uses markers as a point of reference to display the objects. The easiest way to create a marker is to generate a QR code. But markers can also be images which are required to be unique. Images are used in digital newspapers as markers.

Working: After downloading the application, the user has to register and is required to create a unique account. After registration, user is free to use the application according to his/her convenience. The user just requires to scan the image and then the respective object would be superimposed on top of the targeted image. The user would also have the facility for bookmarking the article which he/she would like. The bookmarked article would be saved and the user can view the bookmarked contents whenever he/she wants. The application is very simple to use and thus can be used by a wide variety of users.

Fig. 2 Flow diagram of digital newspaper

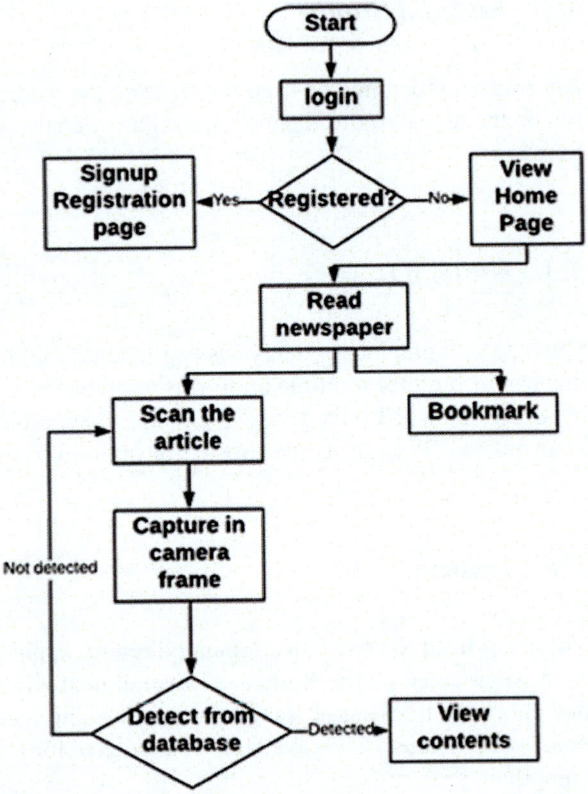

When the user scans the image, the Vuforia SDK compares the image to the images in its database and searches the matching image. Once found, the linked object is displayed in the same plane as that of the target image as long as the target image is in the camera plane of the phone. Thus, the user can view various videos or images associated with the target image on the user's phone. This is explained in Fig. 2.

The admin is required to link all the images with its respective videos, animation, and graphics before the newspaper is out for publication. This process is carried out in the backend part. The backend part is implemented using Vuforia and Unity. Vuforia provides the database required.

4 Implementation Details

In digital newspaper, the user can view various contents related to the scanned image, which is linked at the backend by the newspaper authority. The application intended to digitalize the traditional newspaper. When a particular article is scanned, the user would be able to view various contents related to the scanned image, which would

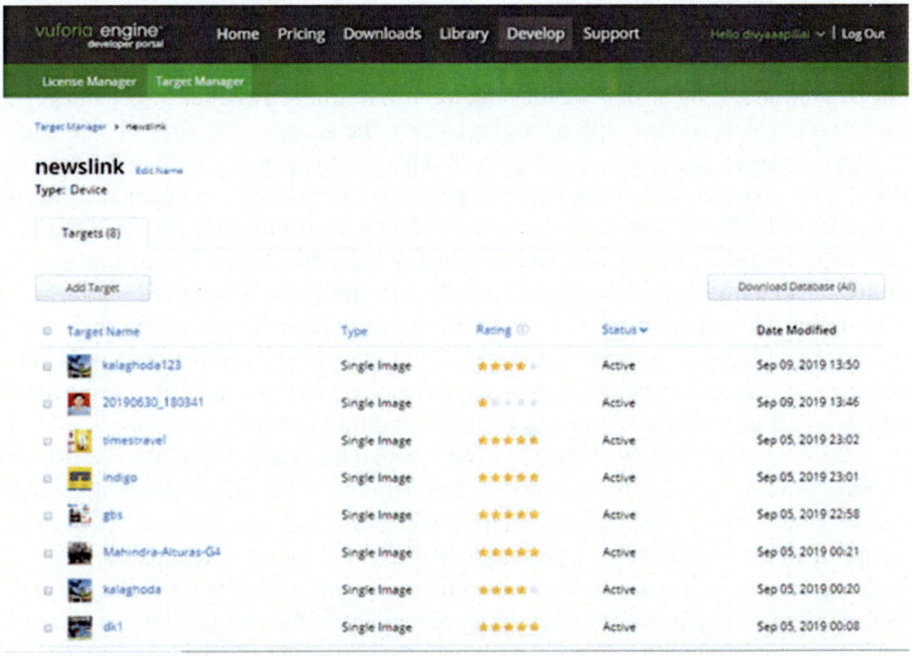

Fig. 3 Database created in Vuforia

be linked at the backend by the newspaper authority. The database would be created in Vuforia which would consist of all the image targets and videos that are required. The database for digital newspaper is shown in Fig. 3.

Digital newspaper is an Android application which consists of various modules.

4.1 URL Redirection

The first module includes redirection to the commercial websites on scanning the advertisement or article. This is done by assigning the targeted image that is the image present in the article using Unity and Vuforia and linking it with the URL of the website to which it is to be redirected. When the user scans the image, the website gets opened on the user's phone.

The target image is stored in the Vuforia database and is assigned using Unity. The button is created in Unity which would appear when the user scans the target image. By clicking the button, the user would be redirected to the linked website.

4.2 Video Playback

The second module consists of the linking article image (marker/target image) to the related objects (image, video, or graphics) in the database. So that when the user scans the image, related object is displayed. This is done with the help of Vuforia and Unity. This involves mostly the backend part of which includes image detection and recognition. In digital newspaper system, Vuforia is used along with Unity2D for displaying respective contents in augmented reality (AR) and Android app development kit for developing Android applications. Vuforia provides a software platform for developing AR applications. Vuforia helps the application to recognize the images and objects which is declared as a marker and helps to perform the declared operation, in this case, it is to display the linked images, videos, and animation. Vuforia acts as a database to store the marker images and linked objects.

In traditional markers like QR code, etc., special black and white lines or patterns are required which would be unique and would uniquely identify the detected image. But Vuforia's image tracker detects and tracks normal images with the help of the natural features found and compares it with different images to identify it uniquely. The targeted image is compared with the images present in the Image Database of the Vuforia and is compared with all the images to identify the image which is currently the target. Once found, the Vuforia will track the image until in the camera's field of view and would display the mentioned object on the tracked image's plane.

Vuforia Target Manager helps to create target images and images could be saved using JPG or PNG format in RGB and Grayscale. Features extracted are then saved in the database, and this database is later used for the comparison while tracking process as runtime comparison. Vuforia gives the application the feature to display AR objects by the scanning process.

4.3 Bookmarking

The fifth module provides bookmarking facilities. So, the user can save the article which he/she likes. The article which is to be saved is also identified uniquely and saved in the database. When the user wants to view them, the system searches its database and displays it. The bookmarked article will be mostly an image in JPG or JPEG format.

5 Result

The Android application that would be created would be having three options to navigate the users, URL redirection, video playback, and bookmarking. The user is free to choose any of the three options.

Fig. 4 Button on the image target

If the user chooses the URL redirection option, then the user is required to scan the image targets present in the newspaper. Different Image Targets would be present throughout the newspaper which would be assigned using Vufoia and Unity by the admin.

Then, the button is assigned using Unity which would be overlapping the current image target when the user scans the image target using their android phone. The button is shown in Fig. 4.

If the user chooses video playback option then again, the procedure of scanning the image targets is same as URL redirection, the only difference is that here the video would be overlaid on the image target and would start playing automatically till the android device is scanning the image target as shown in Fig. 5.

Fig. 5 Video projected on top of image target

6 Conclusion

Once digitization is implemented in newspapers, the readers would benefit on a large scale with more information and knowledge as well as get comfortable with modern technology. The digital newspaper uses Unity and Vuforia to create image targets and to add various functionality related to augmented reality. Vuforia provides a database to store the image targets and associates video, graphics and animation. Android Studio is used to create an android application to facilitate the user to use digital newspapers from their android mobile phones or smartphones. The digital newspaper will also allow the user to save any content they need and access it in the future. With convenient provision for redirecting to online websites of commercial articles and advertisements, it proves to be a user-friendly application and will also reduce the ever-declining demand of newspapers due to the digital media, by providing more opportunities in the advertising field.

The future scope of the project would focus on automating the linking of the videos or URL to the image target automatically which is static now. More focus would be given on using 3D models, so the user would be able to explore more just by one click. Thus, the scope of the project could be increased and this project could be used more efficiently.

References

1. Ling, H.: Augmented reality in reality. IEEE MultiMedia **24**(3), 10–15 (2017)
2. Wilkinson, B., Calder, P.: Augmented reality for the real world. In: International Conference on Computer Graphics, Imaging and Visualisation (CGIV'06). IEEE (2006)
3. Interrante, V., Höllerer, T., Lécuyer, A.: Virtual and augmented reality. IEEE Comput. Graph. Appl. **38**(2), 28–30 (2018)
4. Taketa, N., et al.: Virtual pop-up book based on augmented reality. In: Symposium on Human Interface and the Management of Information. Springer, Berlin, Heidelberg (2007)
5. Valarmathie, G., Zulkifli, A.N., Faisal Mohamed, N.F., Alwi, A., Saidin, A.Z., Mat, R.C., Abu Bakar, J.A.: A review of the features of augmented reality science textbook. In: 1st International Conference on Creative Media, Design & Technology (REKA2014), USM, Penang (2014)
6. Kim, K., Park, J., Woo, W.: Marker-less tracking for multi-layer authoring in AR books. In: International Conference on Entertainment Computing. Springer, Berlin, Heidelberg (2009)
7. Billinghurst, M., Kato, H., Poupyrev, I.: The magic book—moving seamlessly between reality and virtuality. IEEE Comput. Graph. Appl. **21**(3), 2–4 (2001)
8. Kim, S.J.J.: A user study trends in augmented reality and virtual reality research: a qualitative study with the past three years of the ISMAR and IEEE VR conference papers. In: 2012 International Symposium on Ubiquitous Virtual Reality. IEEE (2012)
9. Johnson, L., Levine, A., Smith, R., Stone, S.: Simple augmented reality. The 2010 Horizon Report, 21–24. The New Media Consortium, Austin, TX (2010)
10. Chyan, N.T., Hisham, S.B.: Interactive augmented reality art book to promote Malaysia traditional game. In: 2014 International Conference on Computer, Communications, and Control Technology (I4CT). IEEE (2014)
11. Siltanen, S., et al.: Augmented reality enriches print media and revitalizes media business. Comput. Entertainment (CIE) **15**(3), 4 (2017)
12. Sudarshan, S.K.: Augmented reality in mobile devices (2018)

Nonlocal Fuzzy Solutions for Abstract Second Order Differential Equations

Dimplekumar N. Chalishajar and R. Ramesh

Abstract This work considers the existence and uniqueness of fuzzy solutions for abstract second order differential systems. Since the nonlocal condition has more advantage than local condition, it is studied here. To establish the existence and uniqueness, we apply the concept of semi-group theory and suitable fixed point theorem. Finally, to explain the result, we give an example.

Keywords Mild solution · Fuzzy solution · Fixed point

1 Introduction

Byszewski [4] has studied the existence and uniqueness of strong, classical, and mild solutions of the nonlocal Cauchy problem. The nonlocal condition provides better results in physics when compared with the traditional local conditions. Inspired by this, many authors started to explore this type of equations. See for instance [6, 7], and its application in heat equation is seen in [8] and the references therein. Recently, [13] studied nonlocal conditions in fuzzy metric spaces.

On the other hand, when one is interested in modeling of real-world problems, it is also required sometimes to deal with uncertain phenomena. In this case, the concepts of a fuzzy set are one of the best approach, which leads us to inspect fuzzy differential equations. The implementation of the fuzzy differential equation in the model of population growth is discussed in [18]. Recently, the existence and uniqueness for the solutions to both local and nonlocal conditions for the fuzzy differential equations have been further examined and discussed by several researchers in various aspects, see the monograph of [12] and the papers [2, 3, 5, 9–11, 17]. The role of fuzzy

D. N. Chalishajar (✉)
Department of Applied Mathematics, Virginia Military Institute (VMI), Lexington 24450, USA
e-mail: dipu17370@gmail.com

R. Ramesh
Department of Mathematics, Sri Krishna College of Engineering and Technology, Coimbatore 641008, Tamilnadu, India

O. Castillo et al. (eds.), *Applied Mathematics and Computational Intelligence*, Springer Proceedings in Mathematics & Statistics 413,
https://doi.org/10.1007/978-981-19-8194-4_7

numbers plays a significant role in dealing with vague numerical quantities. For more on fuzzy numbers and its properties, refer [1, 14, 16, 19].

We have tried in this work to describe the existence of fuzzy solutions for the following second order abstract differential system using α techniques of fuzzy numbers.

$$\frac{\mathrm{d}^2}{\mathrm{d}t^2}(p(t) - g(t, p(t))) = Ap(t) + F(t, p(t)), t \in [0, T] = J \tag{1}$$

$$p(0) = \mu(p_0, p), \tag{2}$$

$$\frac{\mathrm{d}}{\mathrm{d}t}(p(t) - g(t, p(t))|_{t=0} = \eta(q_0, p) \tag{3}$$

where $A : [0, T] \to E^n$ is a fuzzy coefficient. The continuous function $g, f : J \times E^n \to E^n$ are nonlinear, $p_0; q_0 \in E^n$ and $\mu, \eta : J \times E^n \to E^n$ are appropriate functions.

So far to the best of our understanding, the existence of solutions for the given differential system (1)–(3) with nonlocal conditions defined in the abstract form is not yet studied using fuzzy techniques, and this serves as a primary motivation for this present work. This work is structured as follows. In Sect. 2, some preliminary concepts of fuzzy sets and fuzzy numbers are provided, and in Sect. 3, the existence and uniqueness of fuzzy solutions are established for the system (1)–(3). Finally, an example is given in Sect. 4 to determine our result.

2 Preliminaries

Here, we review few basic concepts, remarks, and properties of fuzzy numbers which will be used through out this work are presented. For more on its properties refer [15, 19].

Let Q^n be the set of all non-empty compact, convex subsets of $\mathbb{R}^n$. For $C, D \in Q^n$ and for any $\beta \in \mathbb{R}$ the addition and multiplication operation are represented as

$$C + D = \{c + d/c \in C, d \in D\}, C = \{\beta c/c \in C\}$$

In the universe set X, a fuzzy set is defined as the mapping from $m \to [0, 1]$. Here, m is assigned as the degree of membership, and it is value lies between 0 and 1. For the fuzzy set m defined in n-dimensional space and for $\alpha \in (0, 1]$, we denote as,

$$[m]^\alpha = \{x \in R^n/m(x) \geq \alpha\}$$

If m be a fuzzy subset of X, the support of m, denoted as supp(m), is the crisp subset of X whose elements all have nonzero membership values in m, i.e., supp$(m) = \{x \in X|m(x) > 0\}$. For any $\alpha \in [0, 1]$, m is called compact if $[m]^\alpha \in Q^n$.

The collection of all fuzzy sets of R^n is called as E^n which satisfies the following conditions such as m is normal, fuzzy convex, upper semicontinuous, and $[m]^0$ is compact.
For any $m, n \in E^n$ the complete metric H_d is defined as

$$d_\infty(m, n) = \sup_{0<\alpha\leq 1} H_d([m]^\alpha, [n]^\alpha)$$

Let $m, n \in C(J : E^n)$. Then supremum metric is defined as

$$H_1(m, n) = \sup_{0<\alpha\leq T} d_\infty([m]^\alpha, [n]^\alpha)$$

3 Existence and Uniqueness Results

In this section, we define the existence and uniqueness of fuzzy solutions using Banach fixed point theorem for (1)–(3) .
We shall consider a space $\Gamma = p : J \rightarrow E^n$ to define the solution for (1)–(3)
And let us define $\Gamma' = \Gamma \cap C([0, T] : E^n)$

Definition 3.1 A function $p : J \rightarrow E^n$ is an integral solution of (1)–(3), then

$$p(t) = C(t)(\mu(p_0, p) - g(0, \mu(p_0, p))) + K(t)\eta(q_0, p) + g(t, p(t)) \tag{4}$$
$$+ \int_0^t AK(t-s)g(s, p(s))ds + \int_0^t K(t-s)F(s, p(s))ds, t \in J$$

Assume:
(H1) Let $K(t)$ be a fuzzy number, where $[K(t)]^\alpha = [K_l^\alpha(t), K_r^\alpha(t)],\ K(0) = I$ and $K_j^\alpha(t)(j = l, r)$ is continuous with $|K_j^\alpha(t)| \leq m_1,\ m_1 > 0, |AK(t)| \leq m_0, m_0 > 0$ $\forall\, t \in J = [0, T]$.

(H2) Let $C(t)$ be a fuzzy number, where $[C(t)]^\alpha = [C_l^\alpha(t), C_r^\alpha(t)],\ C(0) = I$ and $C_j^\alpha(t)(j = l, r)$ is continuous with $|C_j^\alpha(t)| \leq m_2,\ m_2 > 0\, \forall\, t \in J = [0, T]$.

(H3) $\exists$ positive constants $d_g, d_f > 0$ for the functions g and f which are strongly measurable satisfying the Lipschitz conditions

$$H_d([g(t, p)]^\alpha, [g(t, q)]^\alpha) \leq d_g H_d([p(t)]^\alpha, [q(t)]^\alpha)$$

$$H_d([F(t, p)]^\alpha, [F(t, q)]^\alpha) \leq d_f H_d([p(t)]^\alpha, [q(t)]^\alpha)$$

are satisfied.

(H4) The continuous functions $\mu(p_0, .)$, $\eta(q_0, .)$ are locally bounded and $\exists$ positive constants $d_\mu, d_\eta > 0$ such that

$$H_d([\mu(p_0, p)]^\alpha, [\mu(p_0, q)]^\alpha) \leq d_\mu H_d([p(.)]^\alpha, [q(.)]^\alpha)$$

$$H_d([\eta(q_0, p)]^\alpha, [\eta(q_0, q)]^\alpha) \leq d_\eta H_d([p(.)]^\alpha, [q(.)]^\alpha)$$

(H5) $\left(m_1\left[d_\eta + \frac{d_g}{m_1} + (m_0 d_g + d_f)T\right] + m_2(d_\mu + d_g)\right) < 1$

Theorem 3.1 *Let $T > 0$. If the hypotheses (H1)–(H5) holds, then the system* (1)–(3) *has a unique fuzzy solution in $p \in \Gamma'$.*

Proof For each $p(t) \in \Gamma'$ and $t \in J$, define $(F_0 p)(t) \in \Gamma'$ by,

$$F_0 p(t) = C(t)(\mu(p_0, p) - g(0, \mu(p_0, p))) + K(t)\eta(q_0, p) + g(t, p(t)) + \int_0^t AK(t-s)g(s, p(s))\mathrm{d}s + \int_0^t K(t-s)F(s, p(s))\mathrm{d}s, t \in J.$$

Now,

$$\begin{aligned}
&H_d([F_0 p(t)]^\alpha, [F_0 q(t)]^\alpha)\\
&\leq H_d\Bigg(\Bigg[C(t)(\mu(p_0, p) - g(0, \mu(p_0, p))) + K(t)\eta(q_0, p) + g(t, p(t))\\
&\quad + \int_0^t AK(t-s)g(s, p(s))\mathrm{d}s + \int_0^t K(t-s)F(s, p(s))\mathrm{d}s\Bigg]^\alpha,\\
&\Bigg[C(t)(\mu(p_0, q) - g(0, \mu(p_0, q))) + K(t)\eta(q_0, q) + g(t, q(t))\\
&\quad + \int_0^t AK(t-s)g(s, q(s))\mathrm{d}s + \int_0^t K(t-s)F(s, q(s))\mathrm{d}s\Bigg]^\alpha\Bigg)\\
&\leq H_d\Bigg([C(t)\mu(p_0, p)]^\alpha + [C(t)g(0, \mu(p_0, p))]^\alpha + [K(t)\eta(q_0, p)]^\alpha + [g(t, p(t))]^\alpha\\
&\quad + \left[\int_0^t AK(t-s)g(s, p(s))\mathrm{d}s\right]^\alpha + \left[\int_0^t K(t-s)F(s, p(s))\mathrm{d}s\right]^\alpha,\\
&[C(t)\mu(p_0, q)]^\alpha + [C(t)g(0, \mu(p_0, q))]^\alpha + [K(t)\eta(q_0, q)]^\alpha + [g(t, q(t))]^\alpha\\
&\quad + \left[\int_0^t AK(t-s)g(s, q(s))\mathrm{d}s\right]^\alpha + \left[\int_0^t K(t-s)F(s, q(s))\mathrm{d}s\right]^\alpha\Bigg)\\
&\leq H_d([C(t)\mu(p_0, p)]^\alpha, [C(t)\mu(p_0, q)]^\alpha) + H_d([C(t)g(0, \mu(p_0, p))]^\alpha, [C(t)g(0, \mu(p_0, q))]^\alpha)\\
&\quad + H_d([K(t)\eta(q_0, p)]^\alpha, [K(t)\eta(q_0, q)]^\alpha) + H_d([g(t, p(t))]^\alpha, [g(t, q(t))]^\alpha
\end{aligned}$$

$$
\begin{aligned}
&+ H_d\left(\left[\int_0^t AK(t-s)g(s,p(s))\mathrm{d}s\right]^\alpha, \left[\int_0^t AK(t-s)g(s,q(s))\mathrm{d}s\right]^\alpha\right)\\
&+ H_d\left(\left[\int_0^t K(t-s)F(s,p(s))\mathrm{d}s\right]^\alpha, \left[\int_0^t K(t-s)F(s,q(s))\mathrm{d}s\right]^\alpha\right)\\
&\le m_2 d_\mu H_d([p(.)]^\alpha,[q(.)]^\alpha) + m_2 d_g H_d([p(t)]^\alpha,[q(t)]^\alpha)\\
&+ m_1 d_\eta H_d([p(.)]^\alpha,[q(.)]^\alpha) + d_g H_d([p(t)]^\alpha,[q(t)]^\alpha)\\
&+ m_0 m_1 \int_0^t d_g H_d([p(s)]^\alpha,[q(s)]^\alpha)ds + m_1 \int_0^t d_f H_d([p(s)]^\alpha,[q(s)]^\alpha)\mathrm{d}s
\end{aligned}
$$

Therefore,

$$
\begin{aligned}
&d_\infty(F_0 p(t), F_0 q(t))\\
&= \sup_{0<\alpha\le 1} H_d([F_0 p(t)]^\alpha,[F_0 q(t)]^\alpha)\\
&\le m_2 d_\mu \sup_{0<\alpha\le 1} H_d([p(.)]^\alpha,[q(.)]^\alpha) + m_2 d_g \sup_{0<\alpha\le 1} H_d([p(t)]^\alpha,[q(t)]^\alpha)\\
&\quad + m_1 d_\eta \sup_{0<\alpha\le 1} H_d([p(.)]^\alpha,[q(.)]^\alpha) + d_g \sup_{0<\alpha\le 1} H_d([p(t)]^\alpha,[q(t)]^\alpha)\\
&\quad + m_0 m_1 \int_0^t d_g \sup_{0<\alpha\le 1} H_d([p(s)]^\alpha,[q(s)]^\alpha)\mathrm{d}s\\
&\quad + m_1 \int_0^t d_f \sup_{0<\alpha\le 1} H_d([p(s)]^\alpha,[q(s)]^\alpha)\mathrm{d}s
\end{aligned}
$$

Hence,

$$
\begin{aligned}
H_1(F_0 p, F_0 q) &= \sup_{0\le t\le T} d_\infty(F_0 p(t), F_0 q(t))\\
&\le m_2 d_\mu \sup_{0\le t\le T} d_\infty([p(.)],[q(.)]) + m_2 d_g \sup_{0\le t\le T} d_\infty([p(t)],[q(t)])\\
&\quad + m_1 d_\eta \sup_{0\le t\le T} d_\infty[p(.),[q(.)]) + d_g \sup_{0\le t\le T} d_\infty([p(t)],[q(t)])\\
&\quad + m_0 m_1 d_g \sup_{0\le t\le T} \int_0^t d_\infty([p(s)],[q(s)])\mathrm{d}s\\
&\quad + m_1 d_f \sup_{0\le t\le T} \int_0^t d_\infty([p(s)],[q(s)])\mathrm{d}s\\
&\le \left(m_1\left[d_\eta + \frac{d_g}{m_1} + (m_0 d_g + d_f)T)\right] + m_2(d_\mu + d_g)\right) H_1(p,q)
\end{aligned}
$$

By hypothesis $(H5)$, F_0 is a contraction mapping. Therefore, by Banach fixed point theorem (1)–(3) has a unique solution in $p \in \Gamma'$

4 Example

Here, we establish the results obtained in the previous section to investigate the existence of fuzzy solutions for the wave equation by taking the following second order nonlocal partial differential equation.

$$\frac{\partial}{\partial t}\left[\frac{\partial f(t,\xi)}{\partial t} + \int_0^{\pi} \kappa(\upsilon,\xi) f(t,\upsilon)\mathrm{d}\upsilon\right] = \frac{\partial^2}{\partial \xi^2} f(t,\xi) + G(t, f(t,\xi)), \tag{5}$$

$$f(t,0) = f(t,\pi) = 0, t \in [0,a] \tag{6}$$

$$f(0,\xi) = p(\xi) + \int_0^a m(f(s,\xi))\mathrm{d}s,\ \ \xi \in [0,\pi] \tag{7}$$

$$\frac{\partial}{\partial t} f(0,\xi) = q(\xi) + \int_0^a n(f(s,\xi))\mathrm{d}s \tag{8}$$

Assume:
(i) For the continuous function $G : [0,T] \times [0,\pi] \to \mathbb{R}$, $\exists$ a positive constant $d_G > 0$ such that

$$H_d([G(t,\xi_1)]^{\alpha}, [G(t,\xi_2)]^{\alpha}) \le d_G H_d([p(\xi_1)]^{\alpha}, [q(\xi_2)^{\alpha}]),$$

(ii) The function $m, n : \mathbb{R} \to \mathbb{R}$ are continuous and $\exists$ constants d_m, d_n such that

$$H_d([m(f,\tau_1)]^{\alpha}, [m(f,\tau_2)]^{\alpha}) \le d_m H_d([p(\tau_1)]^{\alpha}, [q(\tau_2)]^{\alpha})$$

$$H_d([n(f,\tau_1)]^{\alpha}, [n(f,\tau_2)]^{\alpha}) \le d_n H_d([p(\tau_1)]^{\alpha}, [q(\tau_2)^{\alpha}])$$

It is evident that by applying the conditions (i) and (ii) in (5)–(8), Theorem (3.1) is satisfied. So, the system (5)–(8) has a unique fuzzy solution.

5 Conclusion

In this work, we have attempted to prove the existence and uniqueness of fuzzy solutions for second order nonlocal abstract differential equations. One can extend the same findings and study the impulsive nature of the systems.

References

1. Alikhani, R., Bahrami, F.: Fuzzy partial differential equations under the cross product of fuzzy numbers. Inf. Sci. **494**, 80–99 (2019)
2. Armand, A., Gouyandeh, Z.: On fuzzy solution for exact second order fuzzy differential equation. Int. J. Ind. Math. **9**(4), 279–288 (2017)
3. Buckley, J.J., Feuring, T.: Fuzzy differential equations. Fuzzy Sets Syst. **110**(1), 43–54 (2000)
4. Byszewski, L.: Theorems about the existence and uniqueness of solutions of a semi-linear evolution nonlocal Cauchy problem. J. Math. Anal. Appl. **162**(2), 494–505 (1991)
5. Chalishajar, D.N., Ramesh, R., Vengataasalam, S., Karthikeyan, K.: Existence of fuzzy solutions for nonlocal impulsive neutral functional differential equations. J. Nonlinear Anal. Appl. **2017**(1), 19–30 (2017)
6. Hernandez, E.: Existence of solutions for an abstract second-order differential equation with nonlocal conditions. Electron. J. Differ. Equ. (EJDE)[electronic only] (2009)
7. Hernández, E., Henríquez, H.: Global solutions for a functional second order abstract Cauchy problem with nonlocal conditions. Annales Polonici Mathematici **2**(83), 149–170 (2004)
8. Ismailov, M.I.: Inverse source problem for heat equation with nonlocal Wentzell boundary condition. Results Math. **73**, 68 (2018). https://doi.org/10.1007/s00025-018-0829-2
9. Kaleva, O.: The Cauchy problem for fuzzy differential equations. Fuzzy Sets Syst. **35**(3), 389–396 (1990)
10. Kumar, M., Kumar, S.: Controllability of impulsive second order semi-linear fuzzy integrodifferential control systems with nonlocal initial conditions. Appl. Soft Comput. **39**, 251–265 (2016)
11. Kwun, Y., Kim, J., Park, M., Park, J.: Nonlocal controllability for the semi-linear fuzzy integrodifferential equations in-dimensional fuzzy vector space. Adv. Differ. Equ. **2009**(1), 734090 (2009)
12. Lakshmikantham, V., Mohapatra, R.N.: Theory of Fuzzy Differential Equations and Inclusions. CRC Press (2004)
13. Long, H.V., Nieto, J.J., Son, N.T.K.: New approach for studying nonlocal problems related to differential systems and partial differential equations in generalized fuzzy metric spaces. Fuzzy Sets Syst. **331**, 26–46 (2018)
14. Mizumoto, M.: Advances in Fuzzy Set Theory and Applications, pp. 153–164. Some properties of fuzzy numbers, Tanaka (1979)
15. Puri, M., Ralescu, D.: Fuzzy random variables. J. Math. Anal. Appl. **114**(2), 409–422 (1986)
16. Qiu, D., Zhang, W., Lu, C.: On fuzzy differential equations in the quotient space of fuzzy numbers. Fuzzy Sets Syst. **295**, 72–98 (2016)
17. Ramesh, R., Vengataasalam, S.: Existence and uniqueness theorem for a solution of fuzzy impulsive differential equations. Ital. J. Pure Appl. Math. **33**, 345–366 (2014)
18. Santo Pedro, F., de Barros, L.C., Esmi, E.: Population growth model via interactive fuzzy differential equation. Inf. Sci. **481**, 160–173 (2019)
19. Wang, G., Li, Y., Wen, C.: On fuzzy n-cell numbers and n-dimensional fuzzy vectors. Fuzzy Sets Syst. **158**(1), 71–84 (2007)

Performance Assessment of Routing Protocols in Cognitive Radio Vehicular Ad Hoc Networks

Jyoti Grover and Sunita Singhal

Abstract Cognitive radio technology is an inventive method to solve spectrum scarcity problems in wireless networks. Growing interest of cognitive radio (CR) technology in vehicular communication systems has led CR-enabled vehicular ad hoc network (CR-VANET). It allows the efficient use of radio frequency spectrum. In vehicular ad hoc networks (VANETs), mobility of nodes leads to frequent changes of network topology making routing a challenging task. It is important to balance different parameters such as impact of spectrum availability, dynamic nature of VANET, varying interference and security while designing solutions for CR-VANETs. In this paper, we have compared AODV and DSR routing protocols in CR-VANET scenario and have found that AODV protocol is more suitable. It is observed that on-demand routing protocols outperform reactive protocols. All the experimental work is conducted through OPNET network simulator.

Keywords Cognitive radio · VANET · CR-VANET · AODV · DSR · Routing

1 Introduction

As the number of vehicles on the road is increasing, it is bringing many challenges such as improving road safety and in-vehicle entertainment. At present, the main objective of vehicle industry is to upgrade the traveling experience of users by strengthening vehicular communication effectiveness with improved safety and performance, as well as Internet connection and commercial applications. Therefore, vehicular ad hoc network (VANET) has developed as an refined technology that can reinforce applications such as safety and traffic monitoring, collision avoidance,

J. Grover (✉)
Malaviya National Institute of Technology, Jaipur, India
e-mail: jgrover.cse@mnit.ac.in

S. Singhal
Manipal University Jaipur, Jaipur, India
e-mail: sunita.singhal@jaipur.manipal.edu

O. Castillo et al. (eds.), *Applied Mathematics and Computational Intelligence*,
Springer Proceedings in Mathematics & Statistics 413,
https://doi.org/10.1007/978-981-19-8194-4_8

multimedia streaming and data collection for smart cities in alliance with wireless sensor networks and vehicle-to-vehicle (V2V) communication.

Government and international companies allocate operation licenses in different frequency bands. The electromagnetic spectrum is overburdened in some bands, while it is under utilized in others. It becomes problematic to assign new licenses as most part of the spectrum is already allocated. On the other hand, some frequency bands are observed with minimum spectrum usage in certain regions of the world. Wireless communication is becoming popular and there is an increased requirement for more bandwidth and electromagnetic spectrum. Additionally, applications like video streaming consume high bandwidth and can cause congestion on particular channel. Spectrum is a scanty resource and it should be used precisely. One of the prominent solutions is to access the spectrum opportunistically. There are more prospects of finding a blank spectrum on highways that can be accessed opportunistically. Thus, it is very useful to embed cognitive radio technology in VANET as spectrum availability can be exploited using opportunistic spectrum access. The main concept of CR network is to share the same bandwidth with unlicensed (secondary) users that was permitted to licensed (primary) users without causing any interference. A CR equipment detects the vacancy of the spectrum and assigns the secondary users to avail the unoccupied radio spectrum.

VANET is egressing technology that allows one hop and multi-hop communications among vehicles in order to build safety applications. Apart from these safety applications, VANET is also used to implement non-safety applications such as multimedia data sharing and Internet access. With the enlarging demand of VANET, cognitive radio is the most promising solution to deal with spectrum scarcity. It is required to provide efficient use of spectrum and proper VANET deployment in order to successfully implement CR-VANET [1]. It acquires geographical environment information to control distribution of spectrum in primary and secondary users so that emergency information can be conveyed timely to nearby vehicles on the road.

Performance of VANET is effected if same communication protocols which were designed for fixed frequency bands are applied in scenario of dynamic use of spectrum bands [2]. Therefore, it is required to develop routing protocols suitable for cognitive radio environment. This paper focuses on the assessment of ad hoc on-demand distance vector (AODV) and dynamic source routing (DSR) in cognitive radio environment. However, there are some papers in literature [3, 4], where different routing protocols are compared in VANET scenario. But we have presented comparative investigation of AODV and DSR routing protocols in CR-VANET. The major contribution of our paper can be summarized as follows:

1. Elaborative description of CR-enabled vehicular ad hoc networks.
2. Assessment of DSR and AODV protocols in cognitive radio environment.
3. Evaluation of applicability of AODV routing protocol in CR-enabled VANET. The organization of the paper is as follows. Section 2 presents background of this technology. In Sect. 3, CR-VANET is revisited. Section 4 presents routing

protocols in CR-enabled VANET. Experimental setup and comparative results are presented in Sect. 5. Finally, concluding remarks with future work are covered in Sect. 6.

2 Background

The main objective of cognitive radio standard is to alleviate the exiguity of spectral resources for wireless communication through smart sensing and rapid resource allocation methods. It is well known that most of the available spectrum is dedicatedly allocated to licensed users (primary users). Generally, it is under utilized depending on the time and the location where communication is taking place. Three approaches have been mentioned below for CR-based communication on the basis of how secondary user accesses the licensed spectrum:

1. Opportunistic spectrum access [5]: In this approach, the secondary users can participate in the communication when a frequency band (mainly used by primary users) is found as idle.
2. Sharing of spectrum [6]: Here, the secondary users synchronize with the primary users and mutually follow a protocol called interference constraint protocol so that quality of service of the primary network can be ensured.
3. Spectrum sharing using sensors [7]: Secondary users sense the status of the channel, i.e., active or idle. Based on the decision made by spectrum sensing, they adapt their transmit power.

A coordinated spectrum sensing framework for cognitive radio VANET is introduced by Xiao et al. [8]. The forwarding node having significant available spectrum bands switches between channels which are common with the receiver and transmitter in order to transmit more data through this forwarding link.

There are some negative effects of cognitive radio technology in vehicular networks. For example, secondary users not detecting a primary (licensed) user and thus causing interference to the primary user. Kakkasageri et al. [9] proposed a model to improve the performance of cognitive radio technology in vehicular networks. It uses cognitive agent model for perceiving intelligent information dissemination. Connectivity-based routing protocols [10] provide spectrum awareness for route selection. It observes the SU network connectivity that is determined by the PU channel usage based upon graph theory and mathematical analysis.

3 Cognitive Radio Vehicular Ad Hoc Network (CR-VANET)

A VANET is a subclass of mobile ad hoc network (MANET) constructed over moving vehicles on the road. There are two types of links formed in VANET:-(a) vehicle-to-vehicle (V2V) and (b) vehicle-to-infrastructure (V2I). It incorporates information and

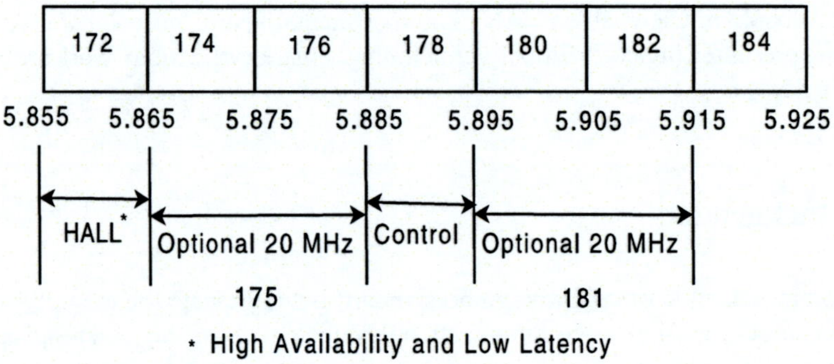

Fig. 1 Dedicated short range communication (DSRC)

communication technology to vehicles and roadside infrastructure. VANET uses 5.9 GHz dedicated short range communication (DSRC) [11, 12] for establishing connection between vehicles and their environment. Currently, IEEE 802.11p standard [13] is working on PHY and MAC layers of DSRC. This standard is an alteration of the popular IEEE 802.11a standard. It deals with the issues such as highly dynamic environment and short span established links. It is a challenge in VANET to complete communication between vehicles in very short time duration due to the high speed of the vehicles. DSRC has been allocated 75 MHz of the radio spectrum by Federal Communications Commission (FCC) to implement vehicular communications. Similar bands are allocated by governments in Japan and Europe. There are 6 service channels and one control channel in 5.9 GHz DSRC spectrum. Bandwidth of each channel is 10 MHz. Different channels of DSRC band are illustrated in Fig. 1.

Each channel is reserved for different types of applications. Safety applications are implemented using channel 172. Channel 174 and 176 are reserved for non-safety applications. Channel 175 is a conjunction of channels 174 and 176. High priority safety messages are sent through channel 178. It supports vehicle-to-vehicle broadcast messages, high priority safety messages and service announcements. Low priority applications are implemented using channel 180 and 182. High priority services are implemented using channel 184.

Wireless technology used for VANET allows nodes to listen to one channel at a time. There is single transceiver equipped in each vehicle according to DSRC deployment. This approach can monitor one channel at a time. Other approach can be used where nodes can be equipped with multiple transceivers. It allows the access of multiple channels simultaneously. This can be understood by taking a simple example. If a vehicle is equipped with two transceivers, one transceiver can monitor the control channel and at the same time send/receive the safety message on a service channel. It becomes expensive and complex by employing multiple transceivers.

Control channel 178 is the most important channel of DSRC. Proper scheduling algorithm is required by the channel for its efficient usage. The control channel is monitored by all the nodes to broadcast safety messages or brief service channel

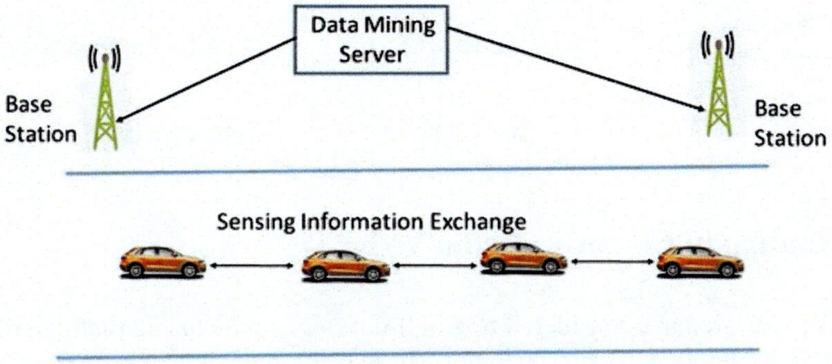

Fig. 2 Spectrum information exchange in cognitive VANET

Fig. 3 Phases of a typical cognitive radio VANET

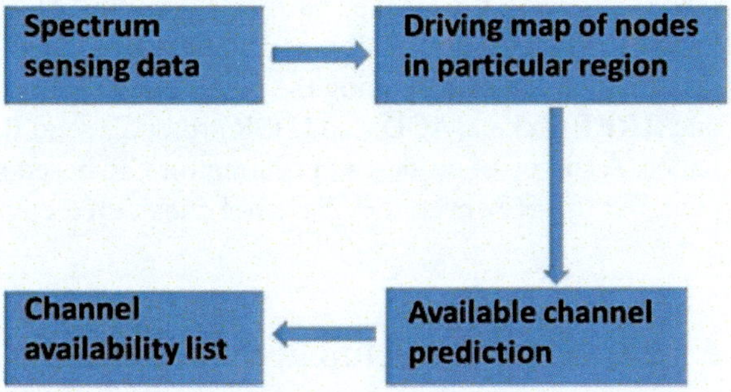

announcements. Communication on the control channel is for a brief duration of time due to availability of limited bandwidth. So, it is required that all the nodes periodically switch to the control channel to receive safety messages. Time synchronization plays a significant role in order to guarantee that safety messages are received by all nodes.

Cognitive radio technology in VANET [14, 15] enables efficient usage of radio spectrum and improves efficiency of vehicular communications. Figure 2 shows a typical spectrum exchange mechanism in cognitive VANET. Vehicles on the road cooperate and exchange spectrum information with each other and nearby base stations. These base stations, further, send the information to data mining servers.

Figure 3 illustrates different phases of a typical cognitive radio VANET. Format of the information maintained by data mining servers is shown in Fig. 4. It stores the list of free channels at each location on highways where free spectrum can be used opportunistically. Cognitive radio VANET improves the performance of current and egressing applications of VANET, e.g., V2V and V2I communication, public safety and non-safety (entertainment, information related) applications.

Fig. 4 Spectrum information maintained by data mining servers

Location Server	Available Channels
$(X_i\ Y_i\ Z_i)$	$(ch_1\ ch_2\ ch_3)$ for user i

4 Routing in Cognitive Radio VANET

In this paper, we have considered two well-developed on-demand routing protocols in VANET: AODV [16] and DSR [17]. In reactive protocols, a source node finds a route to the destination whenever data communication is needed. In this **Route Discovery Phase**, origin (source) node broadcasts a route request (**RREQ**) packet. Nodes receiving this packet, further, forward it. Destination node or any node who knows the path to destination node unicast route reply **RREP** to origin node. Now, source node can start sending the data packets using unicasting. Apart from **RREQ** and **RREP** packets, AODV and DSR protocols send **RERR** packets for route maintenance. A survey of on-demand routing for CR-based networks is presented by Salim et al. in [18]. However, they have not considered cognitive radio in VANET.

5 Experimental Setup and Simulation Results

In our present work, we have assessed AODV and DSR routing protocols in cognitive radio VANET under dynamic traffic scenarios. The simulation model is built in the OPNET simulator [19] with multi-radio multi-channel extensions as it facilitates the simulations of complex networks. The simulation model includes 100 nodes in an area of 3000 m $\times$ 3000 m. We have used two-ray ground propagation model. User Datagram Protocol (UDP) is used at transport layer. Table 1 presents list of parameters for our simulation framework.

Performance of routing protocols in CR-VANET scenario is measured using following parameters:

- **Packet Delivery Ratio (PDR)**: It is the proportion of the number of sent packets by source end and the number of packets collected by the destinations end at application layer.
- **End-to-End Delay**: It is the time difference between sending and receiving a packet from source to destination in a network. It is also called propagation time. This delay is measured in seconds.
- **Throughput**: It is defined as number of packets passing through the network per unit of time. Throughput is measured in Kb/seconds.

Figure5 shows effect of simulation time on end-to-end delay. AODV and DSR routing protocols are considered for evaluation. It is observed that end-to-end delay is more when compared with AODV protocol. This may be due to high speed of

Table 1 Simulation parameters

Parameter	Description
Packet size	1024 bytes
Simulation start time	10 s
Data traffic	Constant bit rate (CBR)
Data rate	2 Mb/s
Simulation area	3000 m × 3000 m
Transmission range	250 m
Simulation time	1000 s
Wireless standard	802.11p
Node speeds	50–100 Km/h
Average speed	1–15 m/s

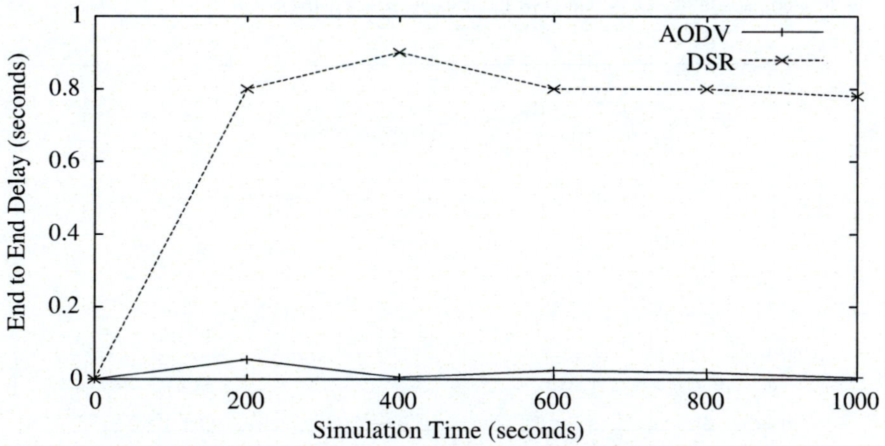

Fig. 5 Comparison of AODV and DSR in terms of end-to-end delay

nodes in VANET. In DSR, **RREQ** and **RREP** packets contain the complete address of all the intermediate nodes. It maintains route cache to store existing routing entries until they become invalid. There are more link failures in VANET, and DSR takes more time to maintain its cache. That is why AODV is having low end-to-end delay as compared to DSR.

PDR is also evaluated by varying the simulation time in Fig. 6. Here, also AODV routing protocol outperforms DSR. This is due to the fact that DSR route discovery phase leads to uncertain length of control and data packets. So, DSR is not suitable because of sporadic connectivity behavior of CR-VANET.

Figure7 considers the effect of simulation time on throughput of both the protocols. Here, also AODV outperforms DSR. This is also due to the same reason as defined above.

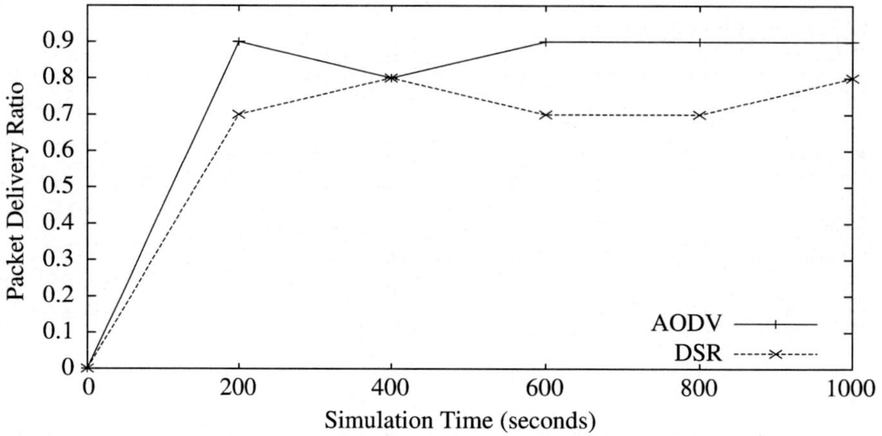

Fig. 6 Comparison of AODV and DSR in terms of packet delivery ratio

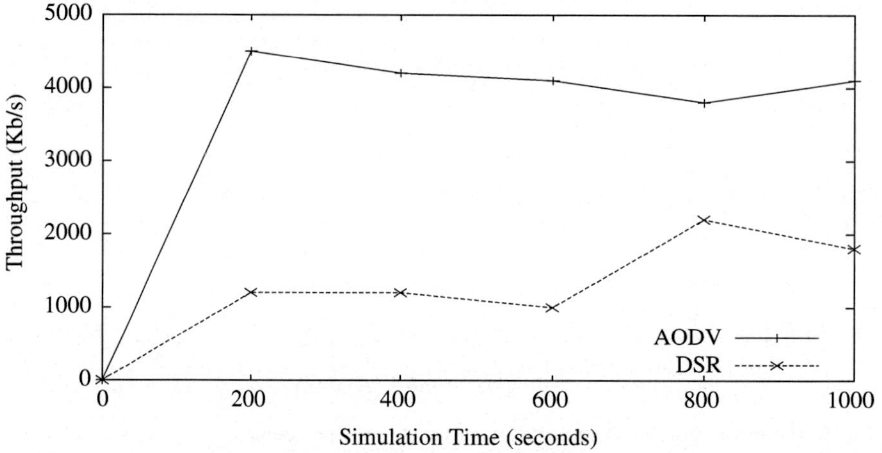

Fig. 7 Comparison of AODV and DSR in terms of throughput

In this paper, we have shown the impact of simulation time on PDR, throughput and end-to-end delay to assess the realization of AODV and DSR routing protocols in CR-VANET.

6 Conclusion and Future Work

This paper discusses the performance assessment of ADOV and DSR routing protocols in CR-VANET. We have not focussed proactive protocols because they have more energy requirements than reactive protocols. Proactive protocols transmit control packet periodically and cache all the paths in a table even if there is not data to be sent. On-demand or reactive protocols behave better in high dynamic scenario such as in VANET. The experimental results illustrate that AODV is more suitable in CR-VANET scenario as compared to DSR. The reason is that DSR route discovery phase causes arbitrary length of control and data packets. Therefore, AODV is found more suitable in CR-VANET. As a part of our future work, we would like to incorporate environmental effects on CR-VANET while assessing the performance of routing protocols.

References

1. Akyildiz, I.F., Lee, W.Y., Chowdhary, K.R.: CRAHNs: cognitive radio ad hoc networks. Ad Hoc Netw. **7**, 810–836 (2009)
2. Cesana, M., Cuomo, F., Ekici, E.: Routing in cognitive radio networks: challenges and solutions. Ad Hoc Netw. **9**(3), 228–248 (2011)
3. Singh, S., Kumari, P., Agrawal, S.: Comparative analysis of various routing protocols in VANET. In: Fifth International Conference on Advanced Computing & Communication Technologies, pp. 315–319. Haryana (2015)
4. Yu, X., Guo, H., Wong, W.: A reliable routing protocol for VANET communication. In: Seventh Wireless communications and Mobile computing conference, pp. 1748–1753. Istanbul, Turkey (2011)
5. Zhang, X., Hang, S.: CREAM-MAC: cognitive radio enabled multi-channel MAC protocol over dynamic spectrum access network. IEEE J. Sel. Top. Signal Process. **5**, 110–123 (2010)
6. Subhedar, M., Birajdar, G.: Spectrum sensing techniques in cognitive radio networks: a survey. Int. J. Next Gener. Netw. **3**, 37–51 (2011)
7. Axell, E., Leus, G., Larsson, E.G., Poor, H.: Spectrum sensing for cognitive radio: state of art and recent advances. IEEE Signal Process. Mag. **29**, 101–116 (2012)
8. Wang, X.Y., Ho, P.H.: A novel sensing coordination framework for CR-VANETs. IEEE Trans. Veh. Technol. **59**, 1936–1948 (2010)
9. Kakkasageri, M.S., Manvi, S.S.: Intelligent information dissemination in vehicular ad hoc networks. Int. J. Ad Hoc Sens. Ubiquitous Comput. **2** (2011)
10. Abbagnale, A., Cuomo, F.: Gymkhana: a connectivity based routing scheme for cognitive radio ad hoc networks. In: IEEE Conference on Computer Communications (INFOCOM). San Diego (2010)
11. Jiang, D., Delgrossi, L.: IEEE 802.11p: towards an international standard for wireless access in vehicular environments. In: Vehicular Technology Conference, pp. 2036–2040 (2008)
12. Eichler, S.: Performance evaluation of the IEEE 802.11p WAVE communication standard. In: IEEE 66th Vehicular Technology Conference, VTC-2007, pp. 2199–2203 (2007)
13. Qing, X., Tony, M., Jeff, K., Sengupta, R.: Vehicle-to-vehicle safety messaging in DSRC. In: 1st ACM International Workshop on Vehicular Ad Hoc Networks, pp. 19–28 (2004)
14. Lim, J.M., Chang, Y.C., Alias, M.Y., Loo, J., Chen, K.: Cognitive radio network in vehicular ad hoc network (VANET): a survey. Cogent Eng. **3**(1), 759–764 (2016)
15. Singh, K.D., Rawat, P., Bonnin, J.: Cognitive radio for vehicular ad hoc networks (CR-VANET): approaches and challenges. EURASIP J. Wireless Commun. Networking **2014**, 49 (2014)
16. Perkins, C., Belding-Royer, E.: Ad-hoc on-demand distance routing protocol. In: 2nd IEEE Workshop on Mobile Computing systems and Applications. New Orleans (1999)
17. Johnson, D.B., Maltz, D.A., Broch, J.: DSR: dynamic source routing for multihop wireless ad hoc networks. Ad Hoc Networking 139–172 (2001)
18. Shelly, S., Sangam, M.: On demand routing protocol for cognitive radio ad hoc networks. EURASIP J. Wireless Commun. Networking **2013**(102) (2013)

19. Chang, X.: Network simulations with OPNET. In: Proceedings of the 31st Conference on Winter Simulation: Simulation—A Bridge to the Future Volume 1 (WSC '99), pp. 307–314 (1999)
20. Cacciapuoti, C.S., Caleffi, M., Paura, L.: Reactive routing for mobile cognitive radio ad hoc networks. Ad Hoc Netw. **10**(5), 803–815 (2012)

Infinite System of Second Order Differential Equations in Banach Space c_0

Tanweer Jalal and Ishfaq Ahmad Malik

Abstract The paper addresses the study of system of second order differential equations of the form:

$$x\frac{d^2 y_j}{dx^2} + \frac{dy_j}{dx} = u_j(x, y(x))$$

where $y_j(1) = y_j(\xi) = 0$, $x \in [1, \xi]$, $v(t) = \left(y_j(x)\right)_{j=1}^{\infty}$ and $j = 1, 2, \ldots$. Using the conception of measure of noncompactness and Darbo's type fixed point theorem, we come up with the condition for the existence of solution of the above system in c_0 space. The result is applied to an example to illustrate the concept.

Keywords Darbo's fixed point theorem · Equicontinuous sets · Infinite system of second order differential equations · Infinite system of integral equations · Measures of noncompactness

1 Introduction

Measures of noncompactness find a great deal of utilization's in operator theory. Mathematicians working in the field of operator theory, functional analysis, and differential equations often use Darbo's theorem [10] to study the fixed point theory. Introduced in 1930 by Kuratowski [19], the idea of measure of noncompactness was broadened to general Banach space by Banaś and Goebel [6]. Banach contraction principle [1, 31] and Schauder's fixed point theorem [18, 20] has been utilized for the study of existence of solutions of functional equations.

T. Jalal (✉) · I. A. Malik
Department of Mathematics, National Institute of Technology Srinagar, Srinagar, India
e-mail: tjalal@nitsri.net

I. A. Malik
e-mail: ishfaq_2phd15@nitsri.net

O. Castillo et al. (eds.), *Applied Mathematics and Computational Intelligence*,
Springer Proceedings in Mathematics & Statistics 413,
https://doi.org/10.1007/978-981-19-8194-4_9

2 Definitions and Preliminaries

Among different measures of noncompactness, the most imperative one is the Hausdorff measure of noncompactness presented by Goldenstein et al. [15] in 1957. The definition and other properties of it can be found in [6].

Let $(M, \|\cdot\|)$ be a Banach space, for any $A \subset X$, $\bar{A}$ denotes closure of A and Conv A denotes the closed convex hull of A. Here, the family of non-empty bounded subsets of M is denoted by $\mathcal{M}_M$ and non-empty; relatively, compact subsets of M are denoted by $\mathcal{N}_M$. Let $\mathbb{R}_+ = [0, \infty)$, the axiomatic definition of measures of noncompactness is

Definition 1 [8] A function $\mu : \mathcal{M}_M \to \mathbb{R}_+$ is said to be the measure of noncompactness on M if the following conditions hold:

(i) The family $\text{Ker}\,\mu = \{A \in \mathcal{M}_M : \mu(A) = 0\}$ is non-empty and $\text{Ker}\,\mu \subset \mathcal{N}_M$;
(ii) $A_1 \subset A_2 \Rightarrow \mu(A_1) \leq \mu(A_2)$;
(iii) $\mu(\bar{A}) = \mu(A)$;
(iv) $\mu(\text{Conv}\,A) = \mu(A)$;
(v) $\mu\,[\lambda A_1 + (1-\lambda)A_2] \leq \lambda\mu(A_1) + (1-\lambda)\mu(A_2)$ for $0 \leq \lambda \leq 1$;
(vi) If (A_n) is a sequence of closed sets from $\mathcal{M}_M$ such that $A_{n+1} \subset A_n$ and $\lim_{n\to\infty} \mu(A_n) = 0$ then the intersection set $A_\infty = \bigcap_{n=1}^{\infty} A_n$ is non-empty.

The fixed point theorem of Darbo type [6, 10] has been used to find the condition for the existence fixed point of an operator which is not compact. It is formulated as follows

Theorem 1 *[10] Let E be a non-empty, bounded, closed, and convex subset of Banach space X and let $T : E \to E$ be a continuous mapping. Assume that there exists a constant $k \in [0, 1)$ such that $\mu\,(T(E)) \leq k\mu(E)$ for any non-empty subset E of X. Then T has a fixed point in the set E.*

Theorem 1 can be applied in a given Banach space M, only if one have formula which expresses the measure of noncompactness in that space. In only few Banach spaces such formulas are known [6, 8].

The c_0 sequence space is the set of all sequences converging to 0. Norm $\|\cdot\|_{c_0}$, on c_0 is defined as

$$\|(x_k)\|_{c_0} = \sup_{k\geq 1}\{|x_k|\}\ ,\quad (x_k) \in c_0.$$

Under the norm $\|\cdot\|_{c_0}$, c_0 is a Banach space, for $E \in \mathcal{M}_{c_0}$, the Hausdorff measure of noncompactness in c_0 is given by

$$\chi_{c_0}(E) = \lim_{n\to\infty}\left\{\sup_{(x_n)\in E}\left(\sup_{n\geq k}|x_n|\right)\right\}. \tag{1}$$

We denote the interval $[1, \xi]$ by J, and consider the Banach space $C(J, M)$ consisting of all functions defined and continuous on $J = [1, \xi]$ with values in a fixed Banach space M. The norm $\|\cdot\|_C$ on $C(J, M)$ is given by

$$\|m\|_C = \sup_{t \in [1,\xi]} \{\|m(t)\|_M\},$$

where $\|\cdot\|_X$ is the norm on X. Now, for the Banach space $C(J, X)$, the Hausdorff measure of noncompactness is given by

$$\chi(E) = \sup_{t \in J} \{\chi_X((E(t))\}, \tag{2}$$

where, $E = \mathcal{M}_{C(J,E)}$, consists of functions that are equicontinuous on J, χ_X is the Hausdorff measure of noncompactness in X and $E(t) = \{h(t) : h \in E\}$. So, for $X = c_0$, and $E \in \mathcal{M}_{C(J,c_0)}$, we have using (1) and (2)

$$\chi(E) = \sup_{t \in J} \left\{ \lim_{k \to \infty} \left\{ \sup_{x_n \in E} \left\{ \sup_{n \geq k} \{|x_n(t)|\} \right\} \right\} \right\}. \tag{3}$$

Lot of work has been done on infinite system of differential equations of order two in different different Banach spaces for references see [3, 7–9, 11, 12, 21–28, 30].

This paper deals with finding existence of solutions of infinite system of differential equations of order 2 of the type

$$x \frac{\mathrm{d}^2 y_j}{\mathrm{d}x^2} + \frac{\mathrm{d}y_j}{\mathrm{d}x} = u_j(x, y(x)) \ ; \quad y_j(1) = y_j(\xi) = 0 \tag{4}$$

where $x \in [1, \xi]$, $y(x) = \left(y_j(x)\right)_{j=1}^{\infty}$.

The study makes use of infinite system of integral equations and Green's function [14]. The system (4) is converted into an equivalent system of integral equations. This system of integral equations is then investigated for the existence of solution in Banach space c_0. The paper ends with an example which supports the idea given.

3 Main Results

A twice differentiable function y_j satisfies (4) if and only if it satisfies the infinite system of integral equations

$$y_j(x) = \int_1^{\xi} X(x, s) u_j(s, y(s)) \mathrm{d}s \ , \quad x \in J \tag{5}$$

where $u_j(x, y) \in C(J, \mathbb{R})$, $j = 1, 2, 3, \ldots$ and the Green's function $X(x, s)$ is given by

$$X(x, s) = \begin{cases} \frac{\ln x(\ln \xi - \ln s)}{\ln \xi}, & 1 \le s < x \le \xi, \\ \\ \frac{\ln s(\ln \xi - \ln x)}{\ln \xi}, & 1 \le x < s \le \xi. \end{cases} \tag{6}$$

This function satisfies the inequality

$$X(x, s) \le 0.25 \ln \xi \tag{7}$$

for all $(x, s) \in J^2$.
Observe that from (5) and (6), we obtain

$$y_j(x) = \int_1^x \frac{\ln x(\ln \xi - \ln s)}{\ln \xi} u_j(s, y(s))ds + \int_x^\xi \frac{\ln(s)(\ln \xi - \ln x)}{\ln(\xi)} u_j(s, y(s))ds.$$

Differentiation this expression, we get

$$\frac{dy_j}{dx} = \frac{1}{\ln \xi} \int_1^x \frac{1}{x}(\ln \xi - \ln s)u_j(s, y(s))ds + \frac{1}{\ln \xi} \int_x^\xi (-\frac{1}{x})(\ln s)u_j(s, y(s))ds$$

$$\Rightarrow x\frac{dy_j}{dx} = \frac{1}{\ln \xi} \int_1^x (\ln \xi - \ln s)u_j(s, y(s))ds - \frac{1}{\ln \xi} \int_x^\xi (\ln s)u_j(s, y(s))ds$$

Again, differentiation gives

$$\begin{aligned} \frac{d}{dx}\left(x\frac{dy_j}{dx}\right) &= \frac{1}{\ln \xi}[0 + (\ln \xi - \ln x)u_j(x, y(x))] - \frac{1}{\ln \xi}[0 - (\ln x)u_j(x, y(x)] \\ &= u_j(x, v(x)). \end{aligned}$$

Since Eq. (4) can be written as $\frac{d}{dx}\left(x\frac{dy_j}{dx}\right) = u_j(x, y(x))$. So, $y_j(t)$ given in (5) satisfies Eq. (4). Hence, the two systems (4) and (5) are equivalent.
In the space c_0 the system (4) has a solution if theses assumptions are satisfied:

(A_1) The functions u_j are real valued, defined on the set $J \times \mathbb{R}^\infty$, $(j = 1, 2, 3, \ldots)$, and the operator u is defined on the space $J \times c_0$ as

$$(x, y) \mapsto (uy)(x) = (u_n(x, y)) = (u_1(x, y), u_2(x, y), u_3(x, y), \ldots)$$

maps the space $J \times c_0$ into c_0. The class of all functions $\{(uy)(x)\}_{x \in J}$ is equicontinuous at every point of the space $y \in c_0$, that is for some $x \in J$, given $\epsilon > 0$, there exists a $\delta > 0$, such that

$$\|(uy_1)(x) - (uy_2)(x)\|_c \leq \epsilon \qquad \text{whenever} \|y_1 - y_2\|_{c_0} \leq \delta$$

for all $y_1, y_2 \in c_0$.
(A_2) For each $x \in J$, $y(x) = \left(y_j(x)\right) \in C(J, c_0)$, the following inequality holds

$$|u_n(x, y(x))| \leq g_n(x) + h_n(x) \sup_{j \geq n} \left\{|y_j|\right\} \qquad n = 1, 2, 3, \ldots$$

where $g_j(x)$ and $h_j(x)$ are non-negative and continuous functions on J. The sequence $\left(g_j(x)\right)_{j\in\mathbb{N}}$ converges uniformly on J to a function vanishing identically on J, and the sequence $\left(h_j(x)\right)_{j\in\mathbb{N}}$ is equibounded on J.

In view of the assumption (A_2), the following constants are finite

$$\mathcal{G} = \sup\{g_n(x) : x \in J, n \in \mathbb{N}\}$$
$$\mathcal{H} = \sup\{h_n(x) : x \in J, n \in \mathbb{N}\}$$

Theorem 2 *Under assumptions (A_1) and (A_2), the infinite system (4) has atleast one solution $y(x) = \left(y_j(x)\right)_{j\in\mathbb{N}}$ in c_0 for some fixed $x \in J$, whenever $(\xi - 1)\mathcal{H} \ln \xi < 4$.*

Proof Define the operator F on $C = C(J, c_0)$ as

$$\begin{aligned}(\mathsf{F}y)(x) &= ((\mathsf{F}y)_n(x)) \\ &= \left(\int_1^{\xi} X(x, s)u_n(s, y_1(s), y_2(s), \ldots)ds\right) \\ &= \left(\int_1^{\xi} X(x, s)u_1(s, y_1(s), y_2(s), \ldots)ds, \int_1^{\xi} X(x, s)u_2(s, y_1(s), y_2(s), \ldots)ds, \ldots\right)\end{aligned}$$

for an arbitrary $y = (y_n) \in C$ and for $x \in J$. First, we show that F maps C into itself. Fix $n \in \mathbb{N}$ and $x \in J$ arbitrarily, then using (7) we have

$$\begin{aligned}|(\mathsf{F}y)_n(x)| &\leq \int_1^{\xi} |X(x, s)||u_n(s, y_1(s), y_2(s), \ldots)|ds \\ &\leq 0.25 \ln \xi \int_1^{\xi} |u_n(s, y_1(s), y_2(s), \ldots)|ds\end{aligned} \tag{8}$$

Now, using assumption (A_1), we obtain

$$\lim_{n\to\infty}\left(0.25\ln\xi\int_1^{\xi}|u_n(s, y_1(s), y_2(s), \ldots)|\mathrm{d}s\right)\to 0$$

Thus, by (8) it is clear that $((\mathsf{F}y)_n(x))\in c_0$, for arbitrarily fixed $x\in J$.
Now, assumptions (A_2) gives

$$\begin{aligned}
|(\mathsf{F}y)_n(x)| &\le \int_1^{\xi}|X(x,s)||u_n(s, y_1(s), y_2(s), \ldots)|\mathrm{d}s\\
&\le 0.25\ln\xi\left[\int_1^{\xi}g_n(s)\mathrm{d}s+\int_1^{\xi}h_n(s)\sup_{j\ge n}\left\{|y_j(s)|\right\}\mathrm{d}s\right]\\
&\le \frac{1}{4}\ln(\xi)\left[(\xi-1)\mathcal{G}+(\xi-1)\mathcal{H}\sup_{x\in J}\left\{\|y(x)\|_{c_0}\right\}\right]\\
&\le \frac{(\xi-1)}{4}\ln\xi[\mathcal{G}+\mathcal{H}\|y\|_C]
\end{aligned}$$

Therefore, the operator F maps C into itself. As $\xi\mathcal{H}\ln\xi<4$, hence the operator $\mathsf{F}: B_r\to B_r$, where B_r is the ball in the space C with radius r and center origin. The radius r is given by

$$r=\frac{(\xi-1)\mathcal{G}\ln\xi}{4-(\xi-1)\mathcal{H}\ln\xi}$$

We prove that F is continuous on B_r, fix $\epsilon>0$ and $y_1\in B_r$ arbitrarily, and take $y_2\in B_r$ such that $\|y_2-y_1\|_{c_0}\le\epsilon$, then for arbitrary $x\in J$ and $n\in\mathbb{N}$ we have

$$\left|(\mathsf{F}y_1)_n(x)-(\mathsf{F}y_2)_n(x)\right|\le\int_1^{\xi}|X(x,s)||u_n(s,y_1)-u_n(s,y_2)|\mathrm{d}s \tag{9}$$

Now, for any $\epsilon>0$, we define the $\delta(\epsilon)$ as

$$\delta(\epsilon)=\sup\left\{|u_n(x,y_1)-u_n(x,y_2)| : y_1, y_2\in c_0, \|y_1-y_2\|_{c_0}\le\epsilon, x\in J, n=1,2,\ldots\right\}.$$

By, assumption (A_1) we see that $\delta(\epsilon)\to$ as $\epsilon\to$. Hence, by (9), we get

$$\|\mathsf{F}y_1-\mathsf{F}y_2\|_C\le\frac{(\xi-1)}{4}\ln\xi\,\delta(\epsilon)$$

Therefore, the the operator F is continuous on B_r. We now show that the set $\mathsf{F}B_r \subset B_r$ is equicontinuous on J. For this fix $\epsilon > 0$ and $y_1 \in B_r$, then for $x_1, x_2 \in J$ with $|x_1 - x_2| \leq \epsilon$, we obtain

$$\left|(\mathsf{F}y)_n(x_1) - (\mathsf{F}y)_n(x_2)\right| \leq \int_1^{\xi} |X(x_1, s) - X(x_2, s)||u_n(s, y)|ds$$

$$\leq \int_1^{\xi} |X(x_1, s) - X(x_2, s)|[g_n(s) + h_n(s) \sup_{j\geq n}\left\{|y_j(s)|\right\}]ds$$

Define,

$$\mathsf{G}(\epsilon) = \sup\left\{|X(x_1, s) - X(x_2, s)| : x_1, x_2, s \in J, |x_1 - x_2| \leq \epsilon\right\}, \tag{10}$$

then we have

$$\|(\mathsf{F}y)(x_1) - (\mathsf{F}y)(x_2)\|_{c_0} \leq (\xi - 1)\mathsf{G}(\epsilon)\left[\mathcal{G} + \mathcal{H}\sup\{\|y(s)\|_{c_0} : s \in J\}\right]. \tag{11}$$

Now, by definition of Green's function $X(t, s)$ it is clear that $\mathsf{G}(\epsilon) \to 0$ as $\epsilon \to 0$.

Therefore, by (11) we conclude that the set $\mathsf{F}B_r$ are equicontinuous on the interval J. Let $Y = \text{Conv}\, \mathsf{F}B_r$, then Y is equicontinuous on J and $Y \subset B_r$. Let Z be a non-empty subset of Y and $y \in Z$. Then, for arbitrarily fixed $x \in J$ and $k \in \mathbb{N}$, we have

$$\left|(\mathsf{F}y)_k(x)\right| \leq \int_1^{\xi} |X(t, s)||u_k(s, y)|ds$$

$$\leq \frac{1}{4}\ln\xi\left(\int_1^{\xi} p_k(s)ds + \int_1^{\xi} q_k(s)\sup_{j\geq k}\left\{|y_j(s)|\right\}ds\right)$$

So, for fixed $n \in \mathbb{N}$ we get

$$\sup_{k\geq n}\left\{\left|(\mathsf{F}y)_k(x)\right|\right\} \leq \frac{1}{4}\ln\xi\left(\int_1^{\xi} \sup_{k\geq n}\{g_k(s)\}ds + \int_1^{\xi} \sup_{k\geq n}\{h_k(s)\}\sup_{k\geq n}\left\{|y_j(s)|\right\}ds\right)$$

Hence, based on assumption (A_2), and taking into account (1), we obtain

$$\chi_{c_0}\left((\mathsf{F}Z)(x)\right) \leq \lim_{n\to\infty}\left\{\sup_{y\in Z}\left\{\sup\left\{\left|(\mathsf{F}y)_k(x)\right| : k \geq n\right\}\right\}\right\}$$
$$\leq \frac{(\xi-1)}{4}\mathcal{H}\ln(\xi)\chi_{c_0}(Z(x)).$$

Therefore, using Theorem (3), we have the inequality

$$\chi(\mathsf{F}Z) \leq \frac{(\xi-1)}{4}\mathcal{H}\ln(\xi)\chi(Z) \tag{12}$$

Hence, by theorem 1, we conclude that the operator F has at least one fixed point y in B_r. The function $y(t) = (y_n(t))$ is a solution of the infinite system of integral equations (5) on J, such that $y(x) \in c_0$ for any $x \in J$.
This implies (4) has a solution in the space c_0.

Note: The value of ξ is chosen in such manner that $(\xi - 1)\mathcal{H}\ln\xi < 4$.
The result is applied to an example to illustrate the concept:

Example 1 Consider the infinite system of differential equations given as

$$t\frac{\mathrm{d}^2 v_j}{\mathrm{d}t^2} + \frac{\mathrm{d}v_j}{\mathrm{d}t} = \frac{t}{n^2}2^{-n^2 t} + \sum_{m=n}^{\infty}\frac{(t^2+1)v_m(t)}{(n^2+1)m^2} \tag{13}$$

with $v_n(1) = v_n(\xi) = 0, \quad t \in [1, \xi], n \in \mathbb{N}$.

Solution: Comparing the given system with (4), we have

$$f_n(t, v) = \frac{t}{n^2}2^{-n^2 t} + \sum_{m=n}^{\infty}\frac{(t^2+1)v_m(t)}{(n^2+1)m^2}.$$

The function $h_{nm}(t) = \dfrac{t^2+1}{(n^2+1)m^2}$ is continuous, and $\sum_{m=n}^{\infty} h_{nm}(t)$ is absolutely uniformly continuous on J, for all $m, n \in \mathbb{N}$.

Using the fact that $h_n(t) = \sum_{m=n}^{\infty}|h_{nm}(t)|$ is uniformly bounded on J, we get

$$\sum_{m=n}^{\infty}\left|\frac{[v_m(t) - u_m(t)](t^2+1)}{(n^2+1)m^2}\right| = \sum_{m=n}^{\infty}\frac{|v_m(t) - u_m(t)|(t^2+1)}{(n^2+1)m^2}$$
$$\leq \|v(t) - u(t)\|\frac{\pi^2}{6}(t^2+1). \tag{14}$$

Now, for fixed $t \in [1, \xi]$, if $v(t) = (v_n(t)) \in c_0$, then $f_n(t, v) \in c_0$ since

$$\begin{aligned}
\lim_{n\to\infty} f_n(t, v) &= \lim_{n\to\infty} \left\{ \frac{t}{n^2} 2^{-n^2 t} + \sum_{m=n}^{\infty} \frac{v_m(t)(t^2+1)}{(n^2+1)m^2} \right\} \\
&\leq \lim_{n\to\infty} \left\{ \frac{t}{n^2 2^{n^2 t}} + \frac{t^2+1}{n^2+1} \sup_{m\geq n} v_m(t) \sum_{m=0}^{\infty} \frac{1}{m^2} \right\} \\
&= \lim_{n\to\infty} \left\{ \frac{t}{n^2 2^{n^2 t}} + \frac{(t^2+1)\pi^2}{6(n^2+1)} \sup_{m\geq n} v_m(t) \right\} \\
&= 0,
\end{aligned}$$

since $(v_m(t)) \to 0$ for each $t \in J$. We now show that (A_1) is satisfied, let $\epsilon > 0$ be arbitrary and $v(t) = (v_n(t)) \in c_0$, then using (14) and taking $u(t) = (u_n(t)) \in c_0$ such that

$$\|v(t) - u(t)\|_{c_0} \leq \delta$$

where δ depends on ϵ , $\delta < \left(\frac{6}{\pi^2(\xi^2+1)}\right)\epsilon$, we get for $t \in J$

$$\begin{aligned}
|f_n(t, v) - f_n(t, u)| &= \left| \sum_{m=n}^{\infty} \frac{[v_m(t) - u_m(t)](t^2+1)}{(n^2+1)m^2} \right| \\
&\leq \sum_{m=n}^{\infty} \frac{|v_m(t) - u_m(t)|(t^2+1)}{(n^2+1)m^2} \\
&\leq \delta \times \frac{\pi^2}{6}(\xi^2+1) \\
&< \epsilon,
\end{aligned}$$

for any fixed n, which implies that $f_n(t, v)$ is continuous, and hence, (A_1) is satisfied. To show that (A_2) is satisfied, we take

$$\begin{aligned}
p_n(t) &= \frac{t}{n^2} 2^{-n^2 t}, \qquad q_n(t) = \frac{\pi^2(t^2+1)}{6(n^2+1)} \\
\mathcal{P} &= \sup\{p_n : n \in \mathbb{N}, t \in J\} = \frac{\xi}{2} \\
\mathcal{Q} &= \sup\{q_n : n \in \mathbb{N}, t \in J\} = \frac{\pi^2(\xi^2+1)}{12}.
\end{aligned}$$

Then for

$$f_n(t, v) = \frac{t}{n^2} 2^{-n^2 t} + \sum_{m=n}^{\infty} \frac{(t^2+1)v_m(t)}{(n^2+1)m^2},$$

we have

$$
\begin{aligned}
|f_n(t, v)| &\leq \left|\frac{t}{n^2}2^{-n^2 t}\right| + \left|\sum_{m=n}^{\infty}\frac{(t^2+1)v_m(t)}{(n^2+1)m^2}\right| \\
&\leq p_n(t) + \frac{(t^2+1)\pi^2}{6(n^2+1)}\sup_{m\geq n}|v_m(t)| \\
&= p_n(t) + q_n(t)\sup_{m\geq n}|v_m(t)|
\end{aligned}
$$

From the above, it is clear that the conditions of (A_2) are verified as well.
Also $Q\xi \tanh(0.5\xi) < 2$ for all $\xi \leq 6.8$.
Therefore, the system given in (13) has a solution in c_0 space, for all $t \in J$.

References

1. Agarwal, R.P., Benchohra, M., Hamani, S.: A survey on existence results for boundary value problems of nonlinear fractional differential equations and inclusions. Acta Appl. Math. **109**(3), 973–1033 (2010)
2. Aghajani, A., Mursaleen, M., Shole Haghighi, A.: Fixed point theorems for Meir-Keeler condensing operators via measure of noncompactness. Acta Math. Sci. **35**(3), 552–566 (2015)
3. Aghajani, A., Pourhadi, E.: Application of measure of noncompactness to ℓ_1-solvability of infinite systems of second order differential equations. Bull. Belg. Math. Soc. Simon Stevin **22**(1), 105–118 (2015)
4. Akhmerov, R.R., Kamenskii, M.I., Potapov, A.S., Rodkina, A.E., Sadovskii, B.N., Appell, J.: Measures of noncompactness and condensing operators. Jahresber. Deutsch. Math.-Verein. **96**(2) (1994)
5. Banaŝ, J.: Applications of measures of weak noncompactness and some classes of operators in the theory of functional equations in the Lebesgue space. Nonlin. Anal. T.M.A. **30**(6) (1997), 3283–3293
6. Banaŝ, J., Goebel, K.: Measures of Noncompactness in Banach Spaces. Lecture Notes in Pure and Appl. Math. Marcel Dekker, New York and Basel (1980)
7. Banaŝ, J., Lecko, M.: Solvability of infinite systems of differential equations in Banach sequence spaces. J. Comput. Appl. Math. **137**(2), (2001), 363–375
8. Banaŝ, J., Mursaleen, M.: Sequence Spaces and Measures of Noncompactness with Applications to Differential and Integral Equations. Springer (2014)
9. Banaŝ, J., Mursaleen, M., Rizvi, S.M.H.: Existence of solutions to a boundary-value problem for an infinite system of differential equations. Electron. J. Diff. Equat. **262**, 1–12 (2017)
10. Darbo, G.: Punti uniti in trasformazioni a codominio non compatto Rend. Sem. Mat. Univ. Padova **24**, 84–92 (1955)
11. Deimling, K.: Ordinary Differential Equations in Banach Spaces. Springer (2006)
12. Deimling, K.: Nonlinear functional analysis. Courier Corporation (2010)
13. Dhage, B.C., O'Regan, D.: A fixed point theorem in Banach algebras with applications to nonlinear integral equation. Funct. Differ. Equ. **7**(3–4), 259–267 (2000)
14. Duffy, G.D.: Green's function with applications. Chapman and Hall/CRC, London (2004)
15. Goldenshtein, L.S., Gokhberg, I.T.S., Markus, A.S.: Investigation of some properties of linear bounded operators in connection with their q-norm. Uchen. Zap. Kishinev. Univ. **29**, 29–36 (1957)

16. Jleli, M., Samet, B.: Existence of positive solutions to a coupled system of fractional differential equations. Math. Method Appl. Sci. **38**, 1014–1031 (2015)
17. Kablin, S., Nibenbebg, L.: On a theorem of P. Nowosad. J. Math. Anal. Appl. **17**, 61–67 (1967)
18. Klamka, J.: Schauder's fixed-point theorem in nonlinear controllability problems. Control Cybernet. **29**, 153–165 (2000)
19. Kuratowski, C.: Sur les espaces complets. Fund. Math. **1**(15), 301–309 (1930)
20. Liu, Z., Shin, M.K.: Applications of Schauder's Fixed-point theorem with respect to iterated functional equations. Appl. Math. Lett. **14**(8), 955–962 (2001)
21. Malik, I.A., Jalal, T.: Application of measure of noncompactness to infinite systems of differential equations in ℓ_p spaces. Rend. Circ. Mat. Palermo II. Ser., 1–12 (2019). https://doi.org/10.1007/s12215-019-00411-6
22. Malik, I.A., Jalal, T.: Boundary value problem for an infinite system of second order differential equations in ℓ_p spaces. Math. Bohem. https://doi.org/10.21136/MB.2019.0086-18
23. Malik, I.A., Jalal, T.: Infinite system of integral equations in two variables of hammerstein type in c_0 and ℓ_1 spaces. Filomat **33**(11), 3441–3455 (2019)
24. Mohiuddine, S.A., Srivastava, H.M., Alotaibi, A.: Application of measures of noncompactness to the infinite system of second order differential equations inℓ_p spaces. Adv. Differ. Equ. **317** (2016). https://doi.org/10.1186/s13662-016-1016-y
25. Mursaleen, M.: Application of measure of noncompactness to infinite system of differential equations. Can. Math. Bull. **56**(2), 388–394 (2013)
26. Mursaleen, M., Mohiuddine, S.A.: Applications of measures of noncompactness to the infinite system of differential equations in ℓ_p spaces. Nonlinear Anal. **75**(4), 2111–2115 (2012)
27. Mursaleen, M., Rizvi, S.M.H.: Existence results for second order linear differential equations in Banach spaces. Appl. Anal. Discrete Math. **12**(2), 481–492 (2018)
28. Mursaleen, M., Rizvi, S.M.H.: Solvability of infinite systems of second order differential equations in c_0 and ℓ_1 by Meir-Keeler condensing operators. Proc. Amer. Math. Soc. **144**(10), 4279–4289 (2016)
29. Muresan, V.: Volterra integral equations with iterations of linear modification of the argument. Novi. Sad. J. Math. **33**, 1–10 (2003)
30. Mursaleen, M., Rizvi, S.M.H., Samet, B.: Solvability of a class of boundary value problems in the space of convergent sequences. Appl. Anal., 1–17 (2017)
31. Olaru, I.M.: An integral equation via weakly Picard operators. Fixed Point Theory **11**(1), 97–106 (2010)

Fourth-Order Computations of Mixed Convection Heat Transfer Past a Flat Plate for Liquid Metals in Elliptical Cylindrical Coordinates

B. Hema Sundar Raju

Abstract The problem of mixed convective heat transfer past a horizontal flat plate for liquid metals has been studied by employing elliptical cylindrical geometry fourth-order compact scheme (ECGFOCS). The coupled fourth-order discretised Navier–Stokes and energy equations are solved by applying a pseudo-time iteration technique. The variation of C_f (skin friction), C_D (drag coefficient), Nu (local Nusselt number), $\overline{\text{Nu}}$ (average Nusselt number) on the plate surface is presented for Reynolds number (Re) from 1 to 200, λ (Richardson number) from 0 to 3.0 and Prandtl number (Pr) of 0.004 (liquid sodium), 0.0208 (liquid gallium alloy) and 0.065 (liquid lithium). It is found that the drag coefficient is increasing with an increase of λ while the same is decreasing with the increase of Re and Pr. It is also found that average Nusselt number is increasing with the increase of Re, λ and Pr. The $C_f\sqrt{\text{Re}}$ values are increasing along the plate surface with an increase of λ, and the same is decreasing with an increase of Pr. At the centre of the plate, the Nu $\text{Re}^{-0.5}$ values are increasing with the increase of λ and Pr. Also, it is verified that as grid refines, the present results approach to fourth-order accuracy.

Keywords Horizontal flat plate · Liquid metals · Mixed convection · Drag coefficient · ECGFOCS · Nusselt number

Nomenclature

λ	Richardson number, $\frac{\text{Gr}}{\text{Re}^2}$
(ξ, η)	Elliptical cylindrical polar coordinates
C_D	Total drag coefficient
C_f	Skin friction coefficient
g	Gravitational acceleration, ms^{-2}
Gr	Grasof number, $\frac{l^3 bg(T_s - T_\infty)}{\nu^2}$
l	Length of the plate, m
Nu	Local Nusselt number
$\overline{\text{Nu}}$	Average Nusselt number

B. Hema Sundar Raju (✉)
Department of Mathematics, National Institute of Technology, Silchar 788010, India
e-mail: drbhsraju@gmail.com; raju@math.nits.ac.in

O. Castillo et al. (eds.), *Applied Mathematics and Computational Intelligence*, Springer Proceedings in Mathematics & Statistics 413,
https://doi.org/10.1007/978-981-19-8194-4_10

Pr Prandtl number, $\frac{\nu}{\alpha}$
Re Reynolds number, $\frac{lU_\infty}{\nu}$
T_∞ Uniform temperature, K
U_∞ Uniform stream velocity, ms^{-1}

Greek Symbols

ψ Dimensionless stream function
ω Dimensionless vorticity
θ Dimensionless temperature
b Thermal coefficient of volumetric expansion, s^{-1}
ν Kinematic viscosity (m^2s^{-1})

Subscripts

∞ Free stream

1 Introduction

The mixed convection heat transfer past a flat plate for liquid metals has been given practical importance due to its engineering and industrial applications. Liquid metals are used as coolant in micro-devices, computer ships because it has many physical properties like higher conductivity, lower density, lower melting point and higher boiling temperature (Li et al. [1]).

The flow past a flat plate has been analysed experimentally by Janour [2] under the presence of the parameter Re from 12 to 2335. Tomotika and Aoi [3] applied Oseen approximation to study the flow over flat plate problem in presence of low parameters, and Janssen [4] has also analysed the same problem by employing electrical analogue technique for Re from 0.1 to 10. Van Dyke [5] modified the correlation formula of drag coefficient which is given by Kuo [6] for the problem of steady flow past a flat plate. Dennis and Dunwoody [7] studied the flat plate problem by using analytical treatment of Oseen's linearised equations for Re from 0.1 to 10^4. Dennis and Chang [8] has applied finite difference method (FDM) to study the steady flow past a flat plate for Re upto 200. Cauchy intergral algorithm is applied by Jobe and Burggraf [9] for solving the asymptotic equations of flow near the trailing edge of flat plate with the presence of Re from 10 to 10^4. Dennis and Smith [10] further updated the work of Dennis and Dunwoody [7] into forced convection heat transfer past a flate plate by similar approach for higher Re and Pr. Vynnycky et al. [11] have studied the conjugate heat transfer problem past a rectangular slab by applying FDM for the presence of high Re and Pr. The forced convection heat transfer past a flat plate in elliptical cylindrical coordinate has been investigated by Juncu [12] by applying FDM for parameters Re from 10 to 400 and Pr from 0.1 to 10. Robertson et al. [13] analysed the mixed convection heat transfer past a flat plate in elliptical cylindrical coordinate by employing FDM for the ranges of parameters Re from 10 to 100, Pr from 0.1 to 10 and λ from 0 to 14.

The numerical schemes applied in the above literature are mainly second-order accurate. For improving accuracy of the numerical scheme Spotz and Carey [14], Li et al. [15], Sanyasiraju and Manjula [16], Raju and Sekhar [17], Sekhar et al. [18–20], Raju et al. [21], Nath et al. [22], Nath and Raju [23] attempted the problem of Navier–Stokes equations in Cartesian, cylindrical and spherical geometry by using fourth-order compact scheme which is stable and fourth-order accurate. Recently, Raju [24] studied the flow (up to Re $= 10^5$) and forced convection heat transfer in liquid metals (Pr $<< 1$) from a flat plate by applying ECGFOCS in pseudo-time iteration technique. The current work aims to study the mixed convection heat transfer past a flat plate for specific liquid metals (liquid sodium (Pr = 0.004), liquid gallium alloy (Pr = 0.0208) and liquid lithium (Pr = 0.065)) by employing ECGFOCS in pseudo-time iteration technique.

2 Problem Formulation

An incompressible, laminar and steady flow of free stream velocity U_∞ and temperature T_∞ past a horizontal flat plate of length l has been considered (as shown in Fig. 1). The gravity vector acts in the opposite direction of y axis of the plate. The non-dimensional stream function, vorticity and energy equations [13] are written in elliptical cylindrical coordinate as follows:
Stream function equation:

$$\frac{\partial^2\psi}{\partial\xi^2}+\frac{\partial^2\psi}{\partial\eta^2}=-Q \tag{1}$$

Vorticity equation:

$$\frac{\partial^2\omega}{\partial\xi^2}+\frac{\partial^2\omega}{\partial\eta^2}=\mathrm{Re}\ \mathrm{J}(\omega,\psi)+\mathrm{R} \tag{2}$$

Energy equation:

$$\frac{\partial^2\theta}{\partial\xi^2}+\frac{\partial^2\theta}{\partial\eta^2}=\mathrm{Pe}\ \mathrm{J}(\theta,\psi) \tag{3}$$

where Jacobian is given by $J(\phi_1,\phi_2)=\frac{\partial\phi_1}{\partial\xi}\frac{\partial\phi_2}{\partial\eta}-\frac{\partial\phi_2}{\partial\xi}\frac{\partial\phi_1}{\partial\eta}$, $Q=\frac{1}{8}(\cosh 2\xi-\cos 2\eta)\,\omega$, $R=-\frac{\lambda\mathrm{Re}}{2}\left(\sinh\xi\cos\eta\frac{\partial\theta}{\partial\xi}-\cosh\xi\sin\eta\frac{\partial\theta}{\partial\eta}\right)$, Pe = Re Pr.

The following boundary conditions are used to solve the Eqs. (1)–(3) as follows.

$$\psi=\frac{\partial\psi}{\partial\xi}=0,\ \omega=-\frac{8}{1-\cos 2\eta}\frac{\partial^2\psi}{\partial\xi^2},\ \theta=1 \text{ at } \xi=0$$

$$\psi\to\frac{1}{2}\sinh\xi\sin\eta,\ \omega=\theta=0 \text{ as } \xi\to\infty$$

$$\psi=\omega=\frac{\partial\theta}{\partial\eta}=0 \text{ at } \eta=0 \text{ and } \eta=\pi$$

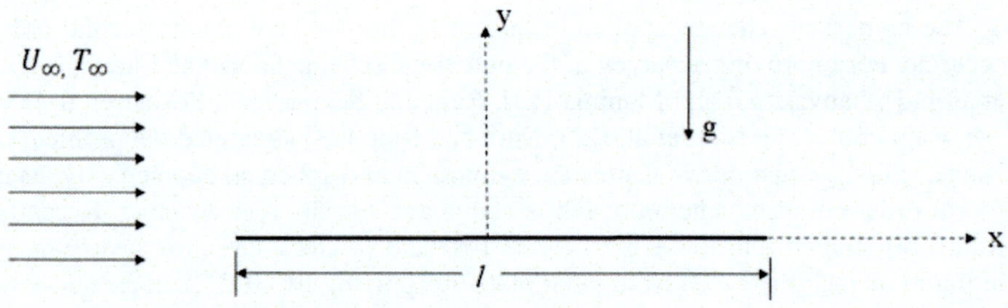

Fig. 1 Schematic diagram describing the problem

The coefficient of drag and skin friction are determined by utilising vorticity field as follows:

$$C_D = -\frac{1}{\text{Re}} \int_0^{\pi} \omega_0 \sin\eta \, d\eta \tag{4}$$

$$C_f = -\frac{(\omega)_{\xi=0}}{\text{Re}} \tag{5}$$

The local and average Nusselt number on the plate surface is computed from temperature field as follows:

$$\text{Nu} = -\frac{2}{\sin\eta}\left(\frac{\partial\theta}{\partial\xi}\right)_{\xi=0} \tag{6}$$

and

$$\overline{\text{Nu}} = -\int_0^{\pi}\left(\frac{\partial\theta}{\partial\xi}\right)_{\xi=0} d\eta \tag{7}$$

3 Numerical Methodology

To discretise the Eqs. (1)–(3), we use the following fourth-order discretizations of the partial derivatives:

$$\frac{\partial Z}{\partial\xi} = Z_\xi - 2f_1\frac{\partial^3 Z}{\partial\xi^3} + O(h^4) \tag{8}$$

$$\frac{\partial^2 Z}{\partial\xi^2} = Z_{\xi\xi} - f_1\frac{\partial^4 Z}{\partial\xi^4} + O(h^4) \tag{9}$$

$$\frac{\partial Z}{\partial\eta} = Z_\eta - 2f_2\frac{\partial^3 Z}{\partial\eta^3} + O(k^4) \tag{10}$$

$$\frac{\partial^2 Z}{\partial \eta^2} = Z_{\eta\eta} - f_2 \frac{\partial^4 Z}{\partial \eta^4} + O(k^4), \tag{11}$$

where Z is a transport variable, $f_1 = \frac{h^2}{12}$ and $f_2 = \frac{k^2}{12}$ with h and k are the step sizes in ξ and η directions, respectively. Z_ξ, Z_η, $Z_{\xi\xi}$ and $Z_{\eta\eta}$ are the second-order central differences. Substitute Eqs. (8)–(11) in Eq. (2), we obtain

$$\begin{aligned} \omega_{\xi\xi} + \omega_{\eta\eta} = {} & f_1 \frac{\partial^4 \omega}{\partial \omega^4} + f_2 \frac{\partial^4 \omega}{\partial \eta^4} \\ & + \mathrm{Re}\left(\psi_\eta \omega_\xi - 2 f_1 \psi_\eta \omega_{\xi\xi\xi} - 2 f_2 \omega_\xi \psi_{\eta\eta\eta} \right. \\ & \left. - \psi_\xi \omega_\eta + 2 f_2 \psi_\xi \omega_{\eta\eta\eta} - 2 f_1 \omega_\eta \psi_{\xi\xi\xi}\right) \\ & - \frac{\lambda \mathrm{Re}}{2} \left[\sinh \xi \cos \eta \left(\theta_\xi - 2 f_1 \theta_{\xi\xi\xi}\right)\right. \\ & \left. - \cosh \xi \sin \eta \left(\theta_\eta - 2 f_2 \theta_{\eta\eta\eta}\right) + O\left(h^4, k^4\right)\right] \end{aligned} \tag{12}$$

By taking the partial derivatives on both sides of vorticity Eq. (2) with respect to ξ and η, we get

$$\frac{\partial^3 \omega}{\partial \xi^3} = -\mathrm{Re}\left(\psi_{\xi\eta} \omega_\xi + \psi_\eta \omega_{\xi\xi} - \psi_{\xi\xi} \omega_\eta - \psi_\xi \omega_{\xi\eta}\right) + R_\xi - \omega_{\xi\eta\eta} \tag{13}$$

$$\begin{aligned} \frac{\partial^4 \omega}{\partial \xi^4} = {} & \mathrm{Re}\left(\psi_{\xi\xi\eta} \omega_\xi + \psi_{\xi\eta} \omega_{\xi\xi} + \psi_{\xi\eta} \omega_{\xi\xi} + \psi_\eta \omega_{\xi\xi\xi} - \omega_\eta \psi_{\xi\xi\xi} \right. \\ & \left. - 2 \psi_{\xi\xi} \omega_{\xi\eta} - \psi_\xi \omega_{\xi\xi\eta}\right) - \omega_{\xi\xi\eta\eta} + R_{\xi\xi} \end{aligned} \tag{14}$$

$$\frac{\partial^3 \psi}{\partial \eta^3} = -\mathrm{Re}\left(\psi_{\eta\eta} \omega_\xi + \psi_\eta \omega_{\xi\xi} - \psi_{\xi\eta} \omega_\eta - \psi_\xi \omega_{\xi\eta}\right) + R_\eta - \omega_{\xi\xi\eta} \tag{15}$$

$$\begin{aligned} \frac{\partial^4 \omega}{\partial \eta^4} = {} & \mathrm{Re}\left(\psi_{\eta\eta\eta} \omega_\xi + \psi_{\eta\eta} \omega_{\xi\eta} + \psi_{\eta\eta} \omega_{\xi\eta} + \psi_\eta \omega_{\xi\eta\eta} - \omega_\eta \psi_{\xi\eta\eta} \right. \\ & \left. - 2 \psi_{\xi\eta} \omega_{\eta\eta} - \psi_\xi \omega_{\eta\eta\eta}\right) - \omega_{\xi\xi\eta\eta} + R_{\eta\eta} \end{aligned} \tag{16}$$

Substitute Eqs. (13)–(16) in Eq. (12), we get

$$\begin{aligned} \omega_{\xi\xi}\left(1 + \beta_1\right) + \omega_{\eta\eta}\left(1 + \gamma_1\right) = {} & \omega_\xi \left(\mathrm{Re}\psi_\eta + \delta_1\right) + \omega_\eta \left(-\mathrm{Re}\psi_\xi + \epsilon_1\right) \\ & + \zeta + O\left(h^4, k^4\right) \end{aligned} \tag{17}$$

The above Eq. (17) is the fourth-order compact discretization form of vorticity equation, where

$$\begin{aligned}
\beta_1 &= f_1 \mathrm{Re}^2 \psi_\eta \psi_\eta - 2 f_1 \mathrm{Re} \psi_{\xi\eta} \\
\gamma_1 &= f_2 \mathrm{Re}^2 \psi_\xi \psi_\xi + 2 f_2 \mathrm{Re} \psi_{\xi\eta} \\
\delta_1 &= (f_1 + f_2)\, \mathrm{Re} \psi_{\xi\xi\eta} - f_1 \mathrm{Re}^2 \psi_\eta \psi_{\xi\eta} + f_2 \mathrm{Re} Q_\eta \omega + f_2 \mathrm{Re}^2 \psi_\xi \psi_{\eta\eta} \\
\epsilon_1 &= -f_1 \mathrm{Re} Q_\xi \omega - (f_1 + f_2)\, \mathrm{Re} \psi_{\xi\eta\eta} + f_1 \mathrm{Re}^2 \psi_\eta \psi_{\xi\xi} - f_2 \mathrm{Re}^2 \psi_\xi \psi_{\xi\eta} \\
\zeta &= -(f_1 - f_2)\, \mathrm{Re} Q \omega_\xi \omega_\eta - 2 f_1 \mathrm{Re} \psi_{\xi\xi} \omega_{\xi\eta} - (f_1 + f_2)\, \mathrm{Re} \psi_\xi \omega_{\xi\xi\eta} \\
&\quad + (f_1 + f_2)\, \mathrm{Re} \psi_\eta \omega_{\xi\eta\eta} + (f_1 + f_2)\, \mathrm{Re}^2 \psi_\xi \psi_\eta \omega_{\xi\eta} - (f_1 + f_2)\, \omega_{\xi\xi\eta\eta} \\
&\quad + 2 f_2 \mathrm{Re} \psi_{\eta\eta} \omega_{\xi\eta} - f_1 \mathrm{Re} \psi_\eta R_\xi + f_2 \mathrm{Re} \psi_\xi R_\eta + f_1 R_{\xi\xi} + f_2 R_{\eta\eta} \\
&\quad - \frac{\lambda \mathrm{Re}}{2} \left[\sinh\xi \cos\eta \left(\theta_\xi - 2 f_1 \theta_{\xi\xi\xi}\right) - \cosh\xi \sin\eta \left(\theta_\eta - 2 f_2 \theta_{\eta\eta\eta}\right) \right]
\end{aligned}$$

The benefit of the present numerical code is that any second-order accurate code can be expanded into fourth-order accurate code by adding the coefficients β_1, γ_1, δ_1, ϵ_1, ζ into the second-order accurate code.

The fourth-order discretised vorticity Eq. (17) is solved by utilizating pseudo-time iterations technique (Erturk and Gökçöl [25]). Apply pseudo-time derivative to the vorticity Eq. (17), we get

$$\begin{aligned}
\omega_t &= \beta \omega_{\xi\xi} + \gamma \omega_{\eta\eta} - \delta \omega_\xi - \epsilon \omega_\eta - \zeta \\
\Rightarrow \frac{\partial \omega}{\partial t} &= \beta \frac{\partial^2 \omega}{\partial \xi^2} + \gamma \frac{\partial^2 \omega}{\partial \eta^2} - \delta \frac{\partial \omega}{\partial \xi} - \epsilon \frac{\partial \omega}{\partial \eta} - \zeta \\
\Rightarrow \frac{\omega^{n+1} - \omega^n}{\Delta t} &= \beta^n \frac{\partial^2 \omega^{n+1}}{\partial \xi^2} + \gamma^n \frac{\partial^2 \omega^{n+1}}{\partial \eta^2} - \delta^n \frac{\partial \omega^{n+1}}{\partial \xi} - \epsilon^n \frac{\partial \omega^{n+1}}{\partial \eta} - \zeta^n \\
\Rightarrow \left(1 - \Delta t \beta^n \frac{\partial^2}{\partial \xi^2} - \Delta t \gamma^n \frac{\partial^2}{\partial \eta^2} + \Delta t \gamma^n \frac{\partial}{\partial \xi} + \Delta t \epsilon^n \frac{\partial}{\partial \eta}\right) \omega^{n+1} &= \omega^n - \Delta t \zeta^n
\end{aligned} \tag{18}$$

Due to the presence of larger band matrix, the above Eq. (18) is not computationally stable and so splits the L.H.S operator of Eq. (18) as follows.

$$\begin{aligned}
&\left(1 - \Delta t \beta^n \frac{\partial^2}{\partial \xi^2} + \Delta t \delta^n \frac{\partial}{\partial \xi}\right)\left(1 - \Delta t \gamma^n \frac{\partial^2}{\partial \eta^2} + \Delta t \epsilon^n \frac{\partial}{\partial \eta}\right) \omega^{n+1} \\
&= \omega^n - \Delta t \zeta^n + \left(\Delta t \beta^n \frac{\partial^2}{\partial \xi^2} - \Delta t \delta^n \frac{\partial}{\partial \xi}\right)\left(\Delta t \gamma^n \frac{\partial^2}{\partial \eta^2} - \Delta t \epsilon^n \frac{\partial}{\partial \eta}\right) \omega^n
\end{aligned} \tag{19}$$

By introducing new variable Y, rearrange the Eq. (19) in two steps (ξ and η sweeps) such that

$$\left(1 - \Delta t \gamma^n \frac{\partial^2}{\partial \eta^2} + \Delta t \epsilon^n \frac{\partial}{\partial \eta}\right) \omega^{n+1} = Y \tag{20}$$

$$\left(1 - \Delta t \beta^n \frac{\partial^2}{\partial \xi^2} + \Delta t \delta^n \frac{\partial}{\partial \xi}\right) Y$$
$$= \omega^n - \Delta t \zeta^n + \left(\Delta t \beta^n \frac{\partial^2}{\partial \xi^2} - \Delta t \delta^n \frac{\partial}{\partial \xi}\right)\left(\Delta t \gamma^n \frac{\partial^2}{\partial \eta^2} - \Delta t \epsilon^n \frac{\partial}{\partial \eta}\right) \omega^n \tag{21}$$

where $\beta = 1 + \beta_1,\ \gamma = 1 + \gamma_1,\ \delta = \mathrm{Re}\psi_\eta + \delta_1,\ \epsilon = -\mathrm{Re}\psi_\xi + \epsilon_1$. The above Eqs. (20)–(21) yields a tridiagonal system of equations and is numerically stable.
By employing TDMA technique, we first solve the Eq. (21) for Y and then solve the Eq. (20) for vorticity (ω). In similar manner, we solve the stream and energy equations.

4 Discretization of Boundary Conditions

Discretization details of boundary conditions are elaborately explained in the papers, Raju and Sekhar [17], Sekhar and Raju [18–20], Raju et al. [21], Nath et al. [22], Nath and Raju [23] and Raju [24].

5 Results and Discussion

To demonstrate the grid sensitivity analysis of ECGFOCS, numerical experiments are carried out in distinct grids, namely 41 × 41, 61 × 61, 81 × 81, 101 × 101, 121 × 121 and 141 × 141 at Re = 200, Pr = 0.0208 and $\lambda = 0.05$. The $\overline{\mathrm{Nu}}$ values attained from the ECGFOCS in the above referred grids are shown in Table 1. The values of $\overline{\mathrm{Nu}}$ referred in Table 1 are almost equal for two grids, namely 121 × 121 and 141 × 141. As a result, for all numerical experiments, 141 × 141 grid has been chosen as an optimum grid size in the current work.

To show the the order convergence of the ECGFOCS, the decay of first-order divided differences of $\overline{Nu}$ values with regard to h are displayed in Fig. 2 together with reference line of fourth order. The slopes of the current results are in line with the fourth-order accurate reference line. It shows that as $h \to 0$, the present results approach to fourth-order accuracy.

The $\overline{\mathrm{Nu}}$ values attained from ECGFOCS are validated with the numerical values of Juncu [12] (for $\lambda = 0$) in Table 2 at Pr = 0.7 and Re = 40, 100 together with the absolute percentage difference. It can be noticed from Table 2 that the present results vary with a maximum percentage difference of 0.0525 %. This indicates that the current results are concurrent with the numerical results of Juncu [12].

Table 1 The $\overline{\text{Nu}}$ values obtained from ECGFOCS for different grid systems at Re = 200, Pr = 0.0208 and λ = 0.05. In bracket, the values indicate step size h of the grid

Grid	$\overline{\text{Nu}}$
41 × 41 (0.07853982)	2.300859
61 × 61 (0.05235988)	2.300120
81 × 81 (0.03926991)	2.299888
101 × 101 (0.03141593)	2.299803
121 × 121 (0.02617994)	2.299766
141 × 141 (0.02243995)	2.299748

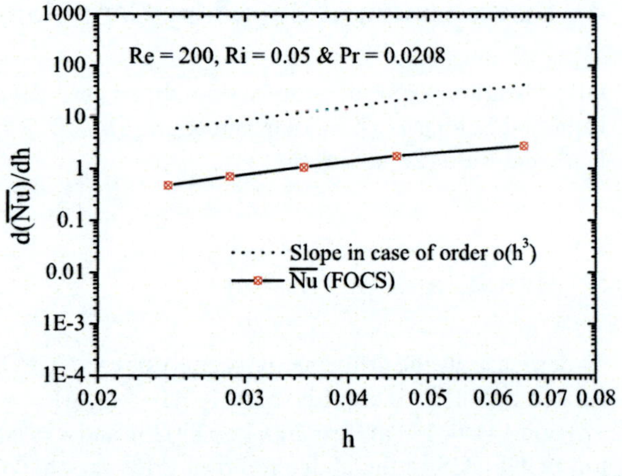

Fig. 2 The variation of $d(\overline{\text{Nu}})/dh$ with refer to step sizes h for Re = 200, λ = 0.05 and Pr = 0.0208

The drag coefficient (C_D) values evaluated from ECGFOCS for distinct values of λ, Re and Pr are plotted in Fig. 3. It is observed from Fig. 3 that for given Pr, the C_D values are increasing with an increase of λ, whereas the same decreases with the increase of Pr and Re.

Figure 4 portrays the variation of $C_f\sqrt{\text{Re}}$ on the plate surface for different values of Pr, λ at Re = 200. It is seen from Fig. 4 that for specific Pr and λ, the $C_f\sqrt{\text{Re}}$ values are first decreasing along the surface of the plate and then increasing near the trailing edge ($x = 0.5$). Again Fig. 4 shows that for particular Pr, the $C_f\sqrt{\text{Re}}$ values are increasing along the plate surface with an increase of λ. Also, it is noticed from Fig. 4 that for a given λ, the $C_f\sqrt{\text{Re}}$ values are decreasing along the surface of the plate with an increase of Pr.

Table 2 Comparison of average Nusselt number obtained from ECGFOCS with the literature values at Pr = 0.7, Re = 40, 100 and $\lambda = 0$ together with the absolute percentage error

Re	Juncu [12]	Present scheme	%Diff
40	4.2800	4.2801	0.0023
100	6.4700	6.4734	0.0525

Figure 5 displays the isothermal contour around the plate for distinct values of Pr and $\lambda = 0, 0.5, 1.5, 3$ at Re = 200. It is seen from Fig. 5 that, the isotherms are densely packed close to the leading edge ($x = -0.5$) of the plate with the increase of λ and Pr. But, the isotherms are more closely packed near the leading edge of the plate with an increase of Pr while compare to increase of a Richardson number.

The variation of Nu $\text{Re}^{-0.5}$ on the plate surface for different values of Pr and λ at Re = 200 is depicted in Fig. 6. It is found that, for given Pr and λ, the Nu $\text{Re}^{-0.5}$ values are first decreasing along the plate surface and then again increasing near the trailing edge. Also, it is observed from Fig. 6 that at the centre of the plate ($x = 0$), the Nu $\text{Re}^{-0.5}$ values are increasing with the increase of λ and Pr.

The variation of $\overline{\text{Nu}}$ with distinct values of λ for Re =10, 100, 200 and Pr = 0.004, 0.0208, 0.065 are presented in Fig. 7. It is seen from Fig. 7 that, $\overline{\text{Nu}}$ is monotonically increased with the increase of λ, Re and Pr.

6 Conclusions

The mixed convection heat transfer past a flat plate has been studied for liquid metals by implementing ECGFOCS in pseudo-time iteration technique. The main points which can be concluded from the current numerical investigations are highlighted as follows.

1. As $h \to 0$, the present results approach to fourth-order accuracy.
2. The drag coefficient values are increasing with an increase of λ, while the same decreases with the increase of Pr and Re.
3. The $C_f\sqrt{\text{Re}}$ values are increasing along the plate surface with an increase of λ, and the same is decreasing with an increase of Pr.
4. The isotherms are more closely packed near the leading edge of the plate with the increase of Pr and λ.
5. At the centre of the plate ($x = 0$), the Nu $\text{Re}^{-0.5}$ values are increasing with the increase of λ and Pr.
6. $\overline{\text{Nu}}$ is monotonically increased with the increase of λ, Re and Pr.

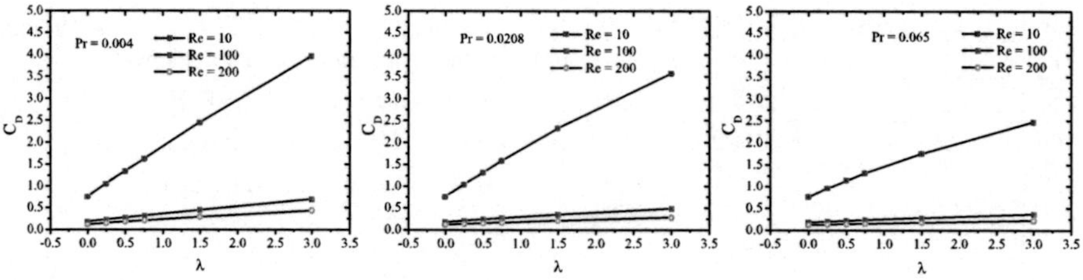

Fig. 3 The C_D values calculated from ECGFOCS for different values of Pr, λ and Re

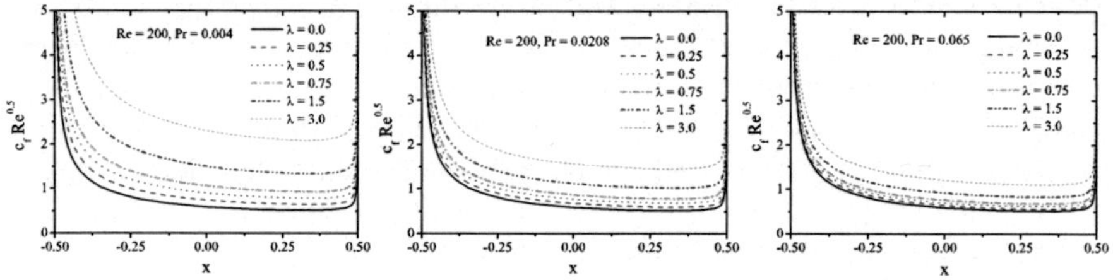

Fig. 4 The variation of $C_f\sqrt{Re}$ on the plate surface for different values of Pr, λ at Re = 200

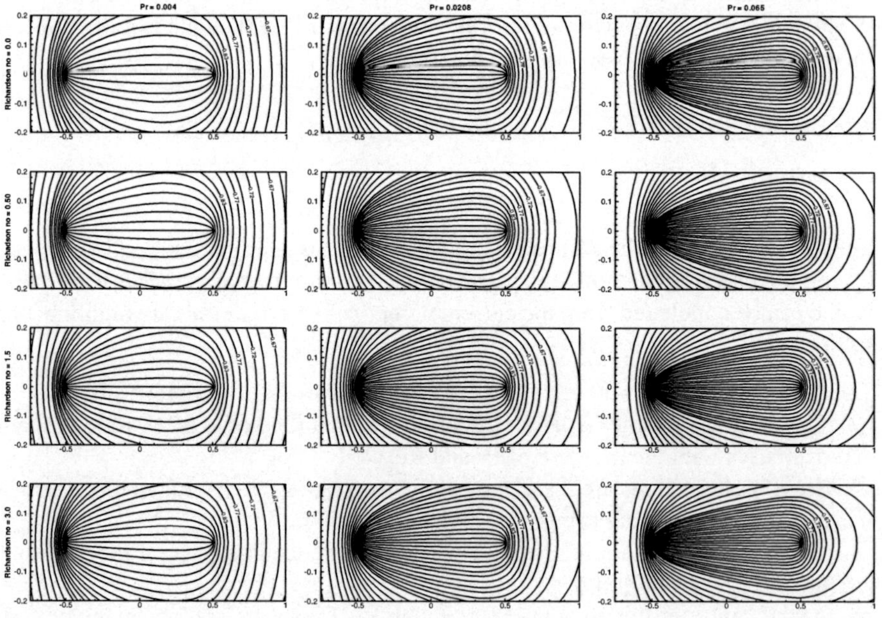

Fig. 5 Isothermal contours for different values of λ = 0, 0.5, 1.5, 3.0 and Pr = 0.004, 0.0208, 0.065 at Re = 200

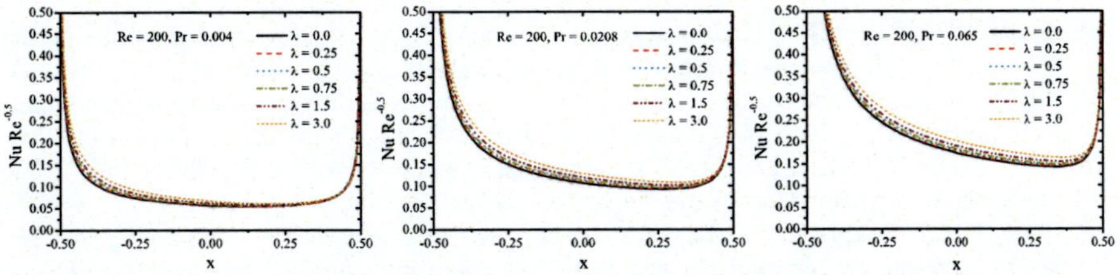

Fig. 6 The variation of Nu $Re^{-0.5}$ on the plate surface for different values of Pr and λ at Re = 200

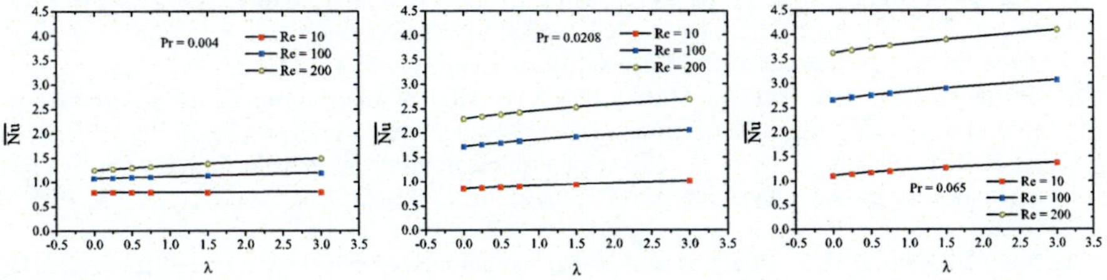

Fig. 7 The $\overline{Nu}$ values evaluted from ECGFOCS with different values of λ for Pr = 0.004, 0.0208, 0.065 and Re = 10, 100, 200

References

1. Li, T., Lv, Y.G., Liu, J., Zhou, Y.X.: A powerful way of cooling computer chip using liquid metal with low melting point as the cooling fluid. Forschung im Ingenieurwesen **70**(4), 243–251 (2005)
2. Janour, Z.: Resistance of a plate in parallel flow at low Reynolds numbers. Nat. Adv. Comm. Aero. Tech. Mem. **1316**, 1–40 (1953)
3. Tomotika, S., Aoi, T.: The steady flow of a viscous fluid past an elliptic cylinder and a flat plate at small Reynolds numbers. Q. J. Mech. Appl. Math. **6**(3), 290–312 (1953)
4. Janssen, E.: Flow past a flat plate at low Reynolds numbers. J. Fluid Mech. **3**(4), 329–343 (1958)
5. Van Dyke, M.: Higher approximations in boundary-layer theory Part 2. Application to leading edges. J. Fluid Mech. **14**, 161–177 (1962)
6. Kuo, Y.H.: On the flow of an incompressible viscous fluid past a flat plate at moderate Reynolds numbers. J. Math. Phys. **32**, 83–101 (1953)
7. Dennis, S.C.R., Dunwoody, J.: The steady flow of a viscous fluid past a flat plate. J. Fluid Mech. **24**(3), 577–595 (1966)
8. Dennis, S.C.R., Chang, G.Z.: Numerical integration of the Navier-Stokes equations for steady two-dimensional flow. Phy. Fluids **12**(12), II–88 (1969)
9. Jobe, C.E., Burggraf, O.R.: The numerical solution of the asymptotic equations of trailing edge flow. R. Soc. A. **340**, 91–111 (1974)
10. Dennis, S.C.R., Smith, N.: Forced convection from a heated flat plate. J. Fluid Mech. **24**(3), 509–519 (1969)
11. Vynnycky, M., Kimura, S., Kanev, K., Pop, I.: Forced convection heat transfer from a flat plate: the conjugate problem. Int. J. heat Mass Trans. **41**(1), 45–59 (1998)
12. Juncu, G.H.: Unsteady forced convection heat/mass transfer from a flat plate. Heat Mass Trans. **41**(12), 1095–1102 (2005)
13. Robertson, G.E., Seinfeld, J.H., Leal, L.G.: Combined forced and free convection flow past a horizontal flat plate. A. I. Ch. E. J. **19**, 998–1008 (1973)

14. Spotz, W.F., Carey. G.F.: High-order compact scheme for the steady stream-function vorticity equations. Int. J. Num. Meth. Eng. **38**, 3497–3512 (1995)
15. Li, M., Tang, T., Fornberg, B.: A compact fourth-order finite difference scheme for the steady incompressible Navier-Stokes equations. Int. J. Num. Meth. Fluids **20**, 1137–1151 (1995)
16. Sanyasiraju, Y.V.S.S., Manjula, V.: Flow past an impulsively started circular cylinder using a higher-order semicompact scheme. Phys. Rev. E **72**, 016709 (2005)
17. Raju, B.H.S., Sekhar, T.V.S.: Higher order compact scheme combined with multigrid method for momentum, pressure poisson and energy equations in cylindrical geometry. Open Numer. Meth. J. **4**, 46–58 (2012)
18. Sekhar, T.V.S., Raju, B.H.S.: Fourth order accurate free convective heat transfer solutions from a circular cylinder. Int. J. Math. Comp. Phys. Electron. Comp. Eng. **6**(8), 1437–1441 (2012)
19. Sekhar, T.V.S., Raju, B.H.S.: An efficient higher order compact scheme to capture heat transfer solutions in spherical geometry. Comput. Phys. Commun. **183**, 2337–2345 (2012)
20. Sekhar, T.V.S., Raju, B.H.S., Murthy, P.V.S.N.: Higher order compact scheme for laminar natural convective heat transfer from a sphere. Appl. Math. Model. **40**(3), 2039–2055 (2016)
21. Raju, B.H.S, Nath, D., Pati, S.: Effect of Prandtl number on thermo-fluidic transport characteristics for mixed convection past a sphere. Int. Commun. Heat Mass Trans. **98**, 191–199 (2018)
22. Nath, D., Raju, B.H.S.: Effect of isoflux thermal boundary condition on mixed convective heat transfer from a sphere for liquid metals. Int. J. Amb. Energy **43**(1), 325–335 (2022)
23. Raju, B.H.S.: Effect of isoflux thermal boundary condition on mixed convective heat transfer from a sphere for liquid metals. Int. J. Amb. Energy (2019). https://doi.org/10.1080/01430750.2019.1636881
24. Raju, B.H.S.: Analysis of forced convection from a horizontal flat plate for liquid metals. Int. J. Amb. Energy **42**(15), 1732–1739 (2021)
25. Erturk, E., Gökçöl, C.: Fourth-order compact formulation of Navier-Stokes equations and driven cavity flow at high Reynolds numbers. Int. J. Num. Meth. Fluids **50**(4), 421–436 (2006)

Steady and Unsteady Solutions of Free Convective Micropolar Fluid Flow Near the Lower Stagnation Point of a Solid Sphere

Debasish Dey and Rupjyoti Borah

Abstract An attempt has been done to study the dual solutions (steady and unsteady solutions) of free convective micropolar fluid flow near the stagnation point of a sphere. Governing equations are solved numerically using suitable similarity transformations and MATLAB built-in bvp4c solver technique. The results are discussed graphically for various values of conjugate parameter (corresponds to convective boundary condition) and material parameter for micropolar fluid. Numerical results of wall temperature and skin friction coefficient are represented by tables. During time-dependent case, skin friction coefficient is controlled by material parameter, but the conjugate parameter enhances skin friction coefficient.

Keywords Convective boundary condition · Lower stagnation point · Micropolar fluid · Solid sphere · Steady and unsteady flow

Nomenclature

$\overline{x}$	Displacement variable along the surface of sphere from lower stagnation point
$\overline{y}$	Displacement variable transverse to $\overline{x}$ axis
$\overline{u}$	Velocity component along $\overline{x}$ axis
$\overline{v}$	Velocity component along $\overline{y}$ axis
$\overline{H}$	Angular velocity of the micropolar fluid
a	Radius of the heated sphere
C_p	Specific pressure
$f'(y)$	Dimensionless velocity
g	Gravitational force
Gr	Grashof number
h_f	Coefficient of heat transfer

D. Dey (✉) · R. Borah
Department of Mathematics, Dibrugarh University, Dibrugarh, Assam, India
e-mail: debasish41092@gmail.com

O. Castillo et al. (eds.), *Applied Mathematics and Computational Intelligence*,
Springer Proceedings in Mathematics & Statistics 413,
https://doi.org/10.1007/978-981-19-8194-4_11

$h(y)$	Dimensionless angular velocity
j	Micro inertia density
k	Vortex viscosity
K	Material or micropolar parameter
k_T	Thermal conductivity
n	Constant
Pr	Prandtl number
T	Temperature
T_f	Temperature of the hot fluid
T_∞	Ambient temperature
t	Time

Greek symbols

β	Coefficient of thermal expansion
ρ	Fluid density
γ_1	Conjugate parameter for convective boundary condition
ψ	Stream function
ϕ	Spin gradient
υ	Kinematic viscosity
μ	Dynamic viscosity
$\theta(y)$	Dimensionless temperature
ω	Eigenvalue parameter

Suffix

0	Initial condition
$'$	Prime which represents the differentiation with respect to y

1 Introduction

The stagnation point flow around a sphere has a large amount of practical significance in engineering and industrial processes. Many researchers have used the mechanics of stagnation point flow using different constitutive models due to its wide range of applications in science and engineering. Theory of micropolar fluid flow is applied in microdevices, defectoscopy (diagnostic method for identification of defects), living organisms, etc. Eringen [1] has introduced the theory of micropolar fluid. Ariman et al. [2] and Rees and Bassom [3] have studied fluid flow problems using micropolar fluid model. Analysis of mixed convective boundary layer flow past a sphere has been done by Chen and Mucoglu [4]. Kadim et al. [5] have investigated the natural convective boundary layer flow past a sphere using a viscoelastic fluid model. Nazar et al. [6, 7] have investigated the micropolar fluid flow past a sphere by taking constant heat flux and wall temperature, respectively. Cheng [8] has investigated the heat and mass transfer effects on micropolar fluid flow past a sphere.

Markin [9] has considered the buoyancy effects on viscous fluid flow with Newtonian heating past vertical plate. Recently, Salleh et al. [10] have investigated the free convection boundary layer flow past a heated sphere using micropolar fluid model. Many researchers (Aziz [11], Makinde and Aziz [12], Ishak et al. [13], Markin and Pop [14], Yao et al. [15], Yacob and Ishak [16]) have carried out the solutions of convective boundary layer flow problems. Recently, Mohamed et al. [17] have studied the behavior of stagnation point flow with convection boundary condition. Stagnation point flow with convective boundary condition using micropolar fluid model has been analyzed by Alkasasbeh[a] et al. [18]. Shu and Wilks [19] have investigated the heat transfer features in the thin-film flow over a hot sphere effecting from a cold vertical plane of liquid falling onto the surface. Aziz et al. [20] have investigated the mixed convection boundary layer flow using viscoelastic micropolar fluid model with the effect of magnetic field.

The main objective of this study is to investigate the steady and unsteady solutions of micropolar fluid flow in the vicinity of the point where velocity is zero of a sphere with convective boundary conditions. The governing partial differential equations are renewed into set of ordinary differential equations using appropriate similarity transformations and have been solved numerically by MATLAB built-in bvp4c solver technique. A comparison of our work has been made with the results of Alkasasbeh[a] et al. [18] to exemplify the truth of the present work.

2 Mathematical Formulation

Here, free convective two-dimensional boundary layer flow has been considered in the region of stagnation point of a heated sphere with the free stream temperature T_∞, which is subjected to a convective boundary condition. Let "a" be the radius of the heated sphere which is shown in Fig. 1.

Under the boundary layer and Boussinesq approximations, the fundamental equations are (following Aziz [11], Eringen [1], Salleh et al. [10], Alkasasbeh[a] et al. [18]):

$$\frac{\partial(\overline{ru})}{\partial\overline{x}} + \frac{\partial(\overline{rv})}{\partial\overline{y}} = 0, \tag{1}$$

$$\rho\left(\overline{u}\frac{\partial\overline{u}}{\partial\overline{x}} + \overline{v}\frac{\partial\overline{u}}{\partial\overline{y}}\right) = (\mu + k)\frac{\partial^2\overline{u}}{\partial\overline{y}^2} + \rho g\beta(T - T_\infty)\sin\left(\frac{\overline{x}}{a}\right) + k\frac{\partial\overline{H}}{\partial\overline{y}}, \tag{2}$$

Fig. 1 Flow model

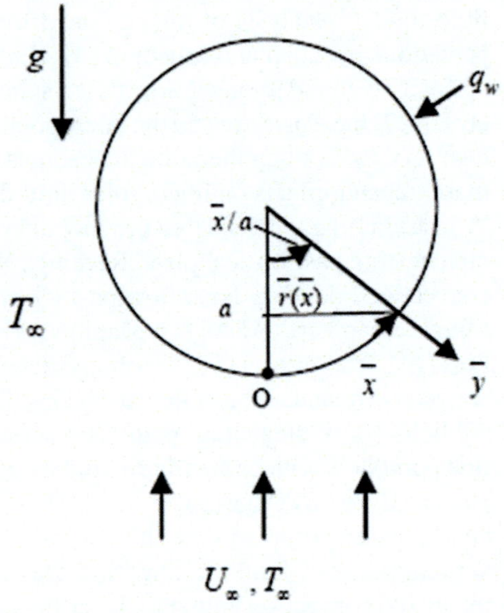

$$\rho j\left(\overline{u}\frac{\partial \overline{H}}{\partial \overline{x}}+\overline{v}\frac{\partial \overline{H}}{\partial \overline{y}}\right)=-k\left(2\overline{H}+\frac{\partial \overline{u}}{\partial \overline{y}}\right)+\phi\frac{\partial^2 \overline{H}}{\partial \overline{y}^2} \tag{3}$$

$$\overline{u}\frac{\partial T}{\partial \overline{x}}+\overline{v}\frac{\partial T}{\partial \overline{y}}=\frac{\upsilon}{\Pr}\frac{\partial^2 T}{\partial \overline{y}^2} \tag{4}$$

and the boundary conditions subject to the above equations are:

$$\left.\begin{array}{l} as\ \ \overline{y}=0:\overline{u}=\overline{v}=0,-k\frac{\partial T}{\partial \overline{y}}=h_f(T_f-T),\overline{H}=-n\frac{\partial \overline{u}}{\partial \overline{y}}, \\ as\ \ \overline{y}\rightarrow\infty:\overline{u}\rightarrow 0,T\rightarrow T_\infty,\overline{H}\rightarrow 0, \end{array}\right\} \tag{5}$$

where $\Pr = \frac{\mu C_p}{k_T}$ and h_f the coefficient of heat transfer for the convective boundary conditions and n the constant with $0 \le n \le 1$ such that the value $n = 0$ which implies $\overline{H} = 0$ at the wall physically signifies the concentrated particle flows in which the density of the particle is significantly large that of microelements near the wall are unable to rotate and $n = 1$ gives turbulent boundary layer (Ahmadi [21]). In this study, we have considered $n = \frac{1}{2}$ which physically corresponds to vanish the antisymmetric part of the stress tensor and gives the weak concentration of microelements.

Let $\overline{r}(\overline{x})$ be the radial distance which characterizes the distance between the surface of the sphere and symmetrical axis of the sphere which is defined by $\overline{r}(\overline{x}) = a\sin(\frac{\overline{x}}{a})$, and ϕ is the spin gradient viscosity that has the following form as proposed by Ahmadi [21]:

$$\phi = (\mu + (\frac{k}{2}))j. \tag{6}$$

Following Salleh et al. [10], Aziz [11], and Alkasasbeh[a] et al. [18], we consider the following dimensionless variables:

$$\left.\begin{array}{l} x = \frac{\bar{x}}{a}, y = Gr^{\frac{1}{4}}(\frac{\bar{y}}{a}), r = \frac{\bar{r}}{a}, u = (\frac{a}{\upsilon})Gr^{-\frac{1}{2}}\bar{u}, v = (\frac{a}{\upsilon})Gr^{-\frac{1}{4}}\bar{v}, \\ H = (\frac{a^2}{\upsilon})Gr^{-\frac{3}{4}}\bar{H}, \theta = \frac{T-T_\infty}{T_f-T_\infty}, \end{array}\right\} \tag{7}$$

where $Gr = \frac{g\beta(T_f-T_\infty)a^3}{\upsilon^2}$ is the Grashof number. Substituting (7) into Eqs. (1–4), we have got

$$\frac{\partial}{\partial x}(ru) + \frac{\partial}{\partial y}(rv) = 0, \tag{8}$$

$$u\frac{\partial u}{\partial x} + v\frac{\partial u}{\partial y} = (1+K)\frac{\partial^2 u}{\partial y^2} + \theta \sin x + K\frac{\partial H}{\partial y}, \tag{9}$$

$$u\frac{\partial H}{\partial x} + v\frac{\partial H}{\partial y} = -K(2H + \frac{\partial u}{\partial y}) + (1 + \frac{K}{2})\frac{\partial^2 H}{\partial y^2}, \tag{10}$$

$$u\frac{\partial \theta}{\partial x} + v\frac{\partial \theta}{\partial y} = \frac{1}{\Pr}\frac{\partial^2 \theta}{\partial y^2}, \tag{11}$$

where $K = \frac{k}{\mu}$, and the reduced boundary conditions are

$$\left.\begin{array}{l} \text{as } y = 0 : u = v = 0, \frac{\partial \theta}{\partial y} = -\gamma_1(1-\theta), H = -\frac{1}{2}\frac{\partial u}{\partial y}, \\ \text{as } y \to \infty : u \to 0, \theta \to 0, H \to 0 \end{array}\right\} \tag{12}$$

where $\gamma_1 = \frac{ah_f Gr^{-\frac{1}{4}}}{k}$. It is seen that if $\gamma_1 = 0$, then $\theta = 0$, and therefore, $h_f = 0$ implies that there is no heat supply from the sphere (Salleh et al. [10] and Alkasasbeh[a] et al. [18]).

We introduce the following similarity variables (following Alkasasbeh[a] et al. [18]):

$$\psi = xr(x)f(x, y), \theta = \theta(x, y), H = xh(x, y) \tag{13}$$

where ψ is the stream function, and $u = \frac{1}{r}\frac{\partial \psi}{\partial y}$ and $v = -\frac{1}{r}\frac{\partial \psi}{\partial x}$ satisfy Eq. (8). Further, we have concentrated our main concern to study the flow behavior in the vicinity of lower stagnation point, and mathematical formulations are done using (following Alkasasbeh et al. [18]) $x \approx 0$. Now, substituting Eq. (13) into Eqs. (9–12), the following ordinary differential equations are obtained:

$$(1+K)f''' + 2ff'' - f'2 + \theta + Kh' = 0, \tag{14}$$

$$(1+\frac{K}{2})h'' + 2fh' - f'h - K(2h + f'') = 0, \tag{15}$$

$$\frac{1}{\Pr}\theta'' + 2f\theta' = 0, \tag{16}$$

and the modernized boundary conditions are:

$$\left.\begin{array}{l} f(0) = f'(0), \theta'(0) = -\gamma_1(1-\theta(0)), h(0) = -\frac{1}{2}f''(0), \\ \text{as } y \to \infty : f'(y) \to 0, \theta(y) \to 0, h(y) \to 0. \end{array}\right\} \tag{17}$$

3 Unsteady Flow Case

The unsteady case of this problem is performed to make a comparison between the steady and unsteady flow solutions and help us to characterize which solution of the dual solutions is physically realizable. Weidman et al. [22] and Khashi'ie et al. [23] have investigated that the dual solutions (steady and unsteady cases) exist for the forced convection boundary layer flow past a permeable flat plate and forced convection flow of a non-Newtonian fluid past a wedge, respectively, and suggested that the upper branch (steady solution) solutions are stable in nature and physically realizable and the lower branch (unsteady solution) solutions are not physically acceptable.

To test these features, we consider the unsteady form of Eqs. (1–4), and (1) clearly holds. Equations (2–4) become.

$$\frac{\partial \overline{u}}{\partial \overline{t}} + \rho\left(\overline{u}\frac{\partial \overline{u}}{\partial \overline{x}} + \overline{v}\frac{\partial \overline{u}}{\partial \overline{y}}\right) = (\mu + k)\frac{\partial^2 \overline{u}}{\partial \overline{y}^2} + \rho g\beta(T - T_\infty)\sin(\frac{\overline{x}}{a}) + k\frac{\partial \overline{H}}{\partial \overline{y}}, \tag{18}$$

$$\frac{\partial \overline{H}}{\partial \overline{t}} + \rho j\left(\overline{u}\frac{\partial \overline{H}}{\partial \overline{x}} + \overline{v}\frac{\partial \overline{H}}{\partial \overline{y}}\right) = -k\left(2\overline{H} + \frac{\partial \overline{u}}{\partial \overline{y}}\right) + \phi\frac{\partial^2 \overline{H}}{\partial \overline{y}^2}, \tag{19}$$

$$\frac{\partial T}{\partial \overline{t}} + \overline{u}\frac{\partial T}{\partial \overline{x}} + \overline{v}\frac{\partial T}{\partial \overline{y}} = \frac{\upsilon}{\Pr}\frac{\partial^2 T}{\partial \overline{y}^2}. \tag{20}$$

We consider the following non-dimensionless variables:

$$\left.\begin{array}{l} t = \frac{Gr^{\frac{1}{2}}}{a^2}\overline{t}, x = \frac{\overline{x}}{a}, y = Gr^{\frac{1}{4}}\left(\frac{\overline{y}}{a}\right), r = \frac{\overline{r}}{a}, u = \left(\frac{a}{\upsilon}\right)Gr^{-\frac{1}{2}}\overline{u}, v = \left(\frac{a}{\upsilon}\right)Gr^{-\frac{1}{4}}\overline{v}, \\ H = \left(\frac{a^2}{\upsilon}\right)Gr^{-\frac{3}{4}}\overline{H}, \theta = \frac{T-T_\infty}{T_f - T_\infty}. \end{array}\right\} \tag{21}$$

Using these dimensionless variables in Eqs. (18–20), we get the following partial differential equations:

$$\frac{\partial u}{\partial t} + u\frac{\partial u}{\partial x} + v\frac{\partial u}{\partial y} = (1+K)\frac{\partial^2 u}{\partial y^2} + \theta \sin x + K\frac{\partial H}{\partial y}, \tag{22}$$

$$\frac{\partial H}{\partial t} + u\frac{\partial H}{\partial x} + v\frac{\partial H}{\partial y} = -K(2H + \frac{\partial u}{\partial y}) + (1+\frac{K}{2})\frac{\partial^2 H}{\partial y^2}, \tag{23}$$

$$\frac{\partial \theta}{\partial t} + u\frac{\partial \theta}{\partial x} + v\frac{\partial \theta}{\partial y} = \frac{1}{\Pr}\frac{\partial^2 \theta}{\partial y^2}, \tag{24}$$

where t denotes time, and it is important for the question of which solution will be obtained physically realizable. Based on the variables (13), we introduce the new dimensionless variables:

$$\psi = xr(x)f(x,y,t), \theta = \theta(x,y,t), H = xh(x,y,t). \tag{25}$$

Using these variables into Eqs. (18–20), we have got the following nonlinear ordinary differential equations:

$$\frac{\partial^2 f}{\partial y \partial t} + x\frac{\partial f}{\partial y}\frac{\partial^2 f}{\partial x \partial y} + \left(\frac{\partial f}{\partial y}\right)^2 - (1 + x\cot x)f\frac{\partial^2 f}{\partial y^2} - x\frac{\partial f}{\partial x}\frac{\partial^2 f}{\partial y^2}$$
$$= (1+K)\frac{\partial^3 f}{\partial y^3} + \frac{\sin x}{x}\theta + K\frac{\partial h}{\partial y}, \tag{26}$$

$$\frac{\partial h}{\partial t} + x\frac{\partial f}{\partial y}\frac{\partial h}{\partial x} + \frac{\partial f}{\partial y}h - (1 + x\cot x)f\frac{\partial h}{\partial y} - x\frac{\partial f}{\partial x}\frac{\partial h}{\partial y}$$
$$= -2Kh - K\frac{\partial^2 f}{\partial y^2} + \left(1 + \frac{K}{2}\right)\frac{\partial^2 h}{\partial y^2}, \tag{27}$$

$$\frac{\partial \theta}{\partial t} + x\frac{\partial f}{\partial y}\frac{\partial \theta}{\partial x} - (1 + x\cot x)f\frac{\partial \theta}{\partial y} - x\frac{\partial f}{\partial x}\frac{\partial \theta}{\partial y} = \frac{1}{\Pr}\frac{\partial^2 \theta}{\partial y^2}. \tag{28}$$

Again, putting $x \approx 0$, Eqs. (26–28) will reduce to the following equations:

$$\frac{\partial^2 f}{\partial y \partial t} + \left(\frac{\partial f}{\partial y}\right)^2 - 2f\frac{\partial^2 f}{\partial y^2} = (1+K)\frac{\partial^3 f}{\partial y^3} + \theta + K\frac{\partial h}{\partial y}, \tag{29}$$

$$\frac{\partial h}{\partial t} + \frac{\partial f}{\partial y}h - 2f\frac{\partial h}{\partial y} = -2Kh - K\frac{\partial^2 f}{\partial y^2} + \left(1 + \frac{K}{2}\right)\frac{\partial^2 h}{\partial y^2}, \tag{30}$$

$$\frac{\partial \theta}{\partial t} - 2f\frac{\partial \theta}{\partial y} = \frac{1}{\Pr}\frac{\partial^2 \theta}{\partial y^2}, \tag{31}$$

and the relevant boundary conditions are:

$$\left.\begin{array}{l} f(0,t) = \frac{\partial f}{\partial y}(0,t) = 0,\ \frac{\partial \theta}{\partial y}(0,t) = -\gamma_1(1-\theta(0,t)),\ h(0,t) = -\frac{1}{2}\frac{\partial^2 f}{\partial y^2}(0,t), \\ \text{as } y \to \infty: \frac{\partial f}{\partial y}(y,t) \to 0, \theta(y,t) \to 0, h(y,t) \to 0. \end{array}\right\} \quad (32)$$

The following representation is taken on to comparison between the dual solutions, according to Weidman et al. [22] and Khashi'ie et al. [23]:

$$\left.\begin{array}{l} f(y,t) = f_0(y) + e^{-\omega t}F(y,t), \\ h(y,t) = h_0(y) + e^{-\omega t}H_1(y,t), \\ \theta(y,t) = \theta_0(y) + e^{-\omega t}G(y,t). \end{array}\right\} \quad (33)$$

where $F(y,t)$, $H_1(y,t)$, and $G(y,t)$ are small relative to the steady flow solutions $f_0(y)$, $h_0(y)$, and $\theta_0(y)$, respectively. The following linearized problems will be obtained by substituting (29) into Eqs. (29–31):

$$(1+K)\frac{\partial^3 F}{\partial y^3} + G(y,t) + K\frac{\partial H_1}{\partial y} + \omega\frac{\partial F}{\partial y} - \frac{\partial^2 F}{\partial y \partial t}$$
$$-2\frac{\partial F}{\partial y} + 2f_0\frac{\partial^2 F}{\partial y^2} + 2F\frac{\partial^2 f_0}{\partial y^2} = 0, \quad (34)$$

$$\left(1+\frac{K}{2}\right)\frac{\partial^2 H_1}{\partial y^2} - K\left[2H_1 + \frac{\partial^2 F}{\partial y^2}\right] + \omega H_1 - \frac{\partial H_1}{\partial t} + \frac{\partial f_0}{\partial y}H_1$$
$$+\frac{\partial F}{\partial y}h_0 + 2f_0\frac{\partial H_1}{\partial y} + 2F\frac{\partial h_0}{\partial y} = 0, \quad (35)$$

$$\frac{1}{\Pr}\frac{\partial^2 G}{\partial y^2} + \omega G - \frac{\partial G}{\partial t} + 2f_0\frac{\partial G}{\partial y} + 2F\frac{\partial \theta_0}{\partial y} = 0 \quad (36)$$

and the boundary conditions are:

$$\left.\begin{array}{l} F(0,t) = \frac{\partial F}{\partial y}(0,t) = 0,\ \frac{\partial G}{\partial y}(0,t) = \gamma_1 G(0,t),\ H_1(0,t) = -\frac{1}{2}\frac{\partial^2 F}{\partial y^2}(0,t), \\ \text{as } y \to \infty: \frac{\partial F}{\partial y}(y,t) \to 0, G(y,t) \to 0, H_1(y,t) \to 0. \end{array}\right\}. \quad (37)$$

The solutions $f(y) = f_0(y), h(y) = h_0(y)$ and $\theta(y) = \theta_0(y)$ of the steady Eqs. (14–16) are obtained by setting $t = 0$. The function $F(y) = F_0(y)$, $H_1(y) = H_{1_0}(y)$, and $G(y) = G_0(y)$ will identify the initial growth or decay of disturbances of the solutions of Eqs. (34–36). Thus, the linearized eigenvalue problems are given by

$$(1+K)F_0''' + G_0 + KH_{1_0}' + (\omega - 2)F_0' + 2\big(f_0F_0'' + F_0f_0''\big) = 0, \quad (38)$$

$$\left(1+\frac{K}{2}\right)H_{1_0}+(f_0'-2+\omega)H_{1_0}-KF_0+F_0'+2f_0H_{1_0}+2F_0=0, \quad (39)$$

$$\frac{1}{\Pr}G_0+\omega G_0+2\left(f_0G_0'+F_0\theta_0'\right)=0 \quad (40)$$

Related boundary conditions are:

$$\left.\begin{array}{c} F_0(0)=F_0'(0)=0,\ G_0'(0)=\gamma_1G_0(0),\ H(0)=-\frac{1}{2}F_0''(0) \\ \text{as } \infty : F(0)\to 0,\ G_0(0)\to 0,\ H_0(0)\to 0 \end{array}\right\} \quad (41)$$

The nature of the steady and unsteady flow solutions depends on the smallest eigenvalue, ω_1. Following Harish et al. [24], we reduce the boundary condition $F_0'(\infty)\to 0$ to a new boundary condition, $F_0(0)=1$, to evaluate the fixed value of eigenvalues and hence solve (38–40).

4 Results and Discussion

The outcome of this work highlights the effect of two parameters, namely the material parameter K and the conjugate parameter γ_1 (parameter responsible for convective boundary condition) on the velocity, angular velocity, and temperature profiles. Here, we have considered $x \approx 0$ (neighborhood of lower stagnation point of the sphere), and the Prandtl number Pr is fixed to 0.7 (Pr $\ll$ 1), which physically indicates the liquid metals-air, which have high thermal conductivity) throughout this paper. We have considered the values of the material parameter $K = 0, 1, 2$, i.e., $K = 0$ characterizes Newtonian fluid and nonzero values signify micropolar fluids. Also, we have considered the conjugate parameter $\gamma_1 \leq 1$, which physically indicates the higher vortex viscosity fluid. All the flow profiles (2–6) have satisfied the convective boundary conditions asymptotically. The visualizations of steady and unsteady flow are emphasized on Figs. 2, 3, 4, 5 and 6. The first solution (steady solution) is denoted by a solid line, and the second solution (unsteady solution) is represented by the dashed line.

Tables 1and 2 represent the numerical results of the first and second solutions of the skin friction coefficient $f(0)$ and wall temperature $\theta(0)$ for $K = 0, 1, 2$ when $\Pr = 0.7$ and $\gamma_1 = 0.05, 0.2$. We have compared our first solutions (steady flow) of skin friction coefficient and wall temperature with the work of Alkasasbeh[a] et al. [18] in Tables 1 and 2, respectively.

From these tables, it is seen that for a fixed value of γ_1, skin friction coefficient for time-independent flow experiences reduction in the magnitude with the enhancement of K, but a reverse pattern is experienced during time-dependent fluid flow around a solid sphere. Further, the convective boundary condition helps to magnify skin friction coefficient. The skin friction coefficient of Newtonian fluid flow is more than micropolar fluid flow (steady case), but opposite behavior is observed for unsteady

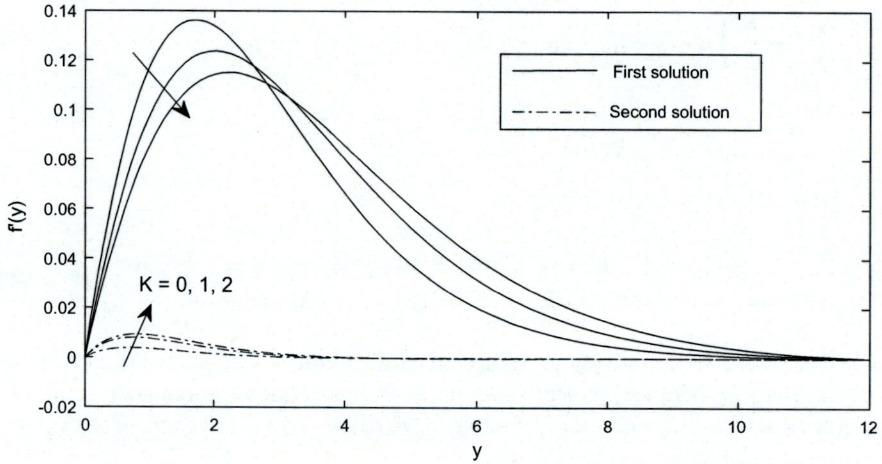

Fig. 2 Velocity distributions $f'(y)$ for some values of K when $\mathrm{Pr} = 0.7$ and $\gamma_1 = 0.05$

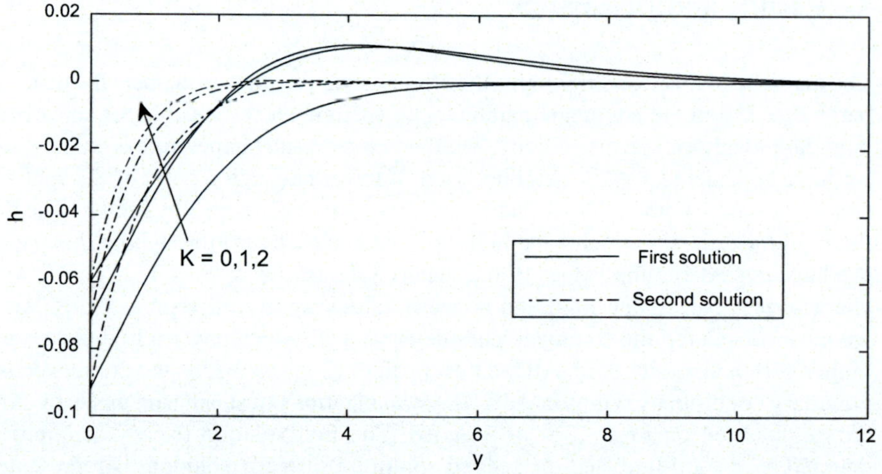

Fig. 3 Angular velocity distributions $h(y)$ for some values of K when $\mathrm{Pr} = 0.7$ and $\gamma_1 = 0.05$

cases. Similarly, temperature of fluid at the surface experiences reduction with K, and it is maximum for Newtonian fluid than micropolar fluids during time-independent cases, but during time-dependent cases, K helps to enhance the wall temperature.

Figures 2 and 3 characterize the influence of material parameter K for $\gamma_1 = 0.05$ and $\mathrm{Pr} = 0.7$ on the velocity and angular velocity in the neighborhood of lower stagnation point of the sphere, $x \approx 0$. Fluid reaches its maximum velocity during $K = 0$ (Newtonian case) and then gradually it decreases because of the presence of vortex viscosity. Small variation between Newtonian and non-Newtonian cases is seen during unsteady fluid flows.

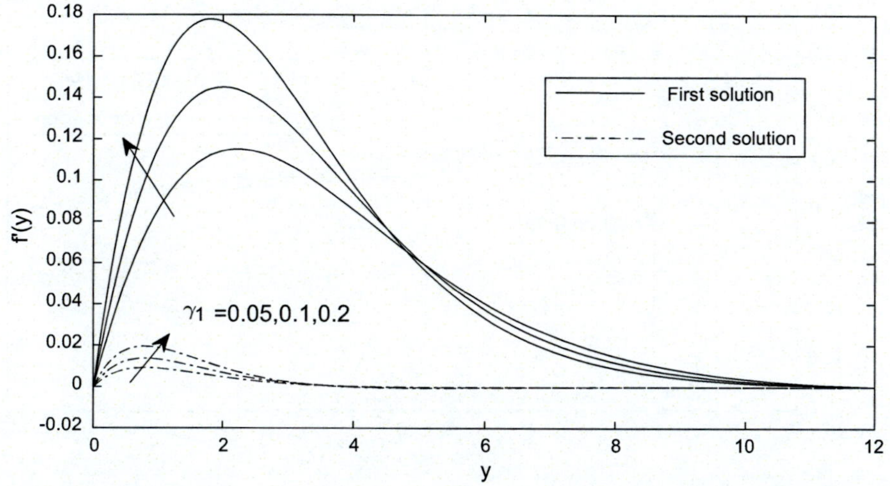

Fig. 4 Velocity distributions $f'(y)$ for some values of γ_1 when $\Pr = 0.7$ and $K = 2$

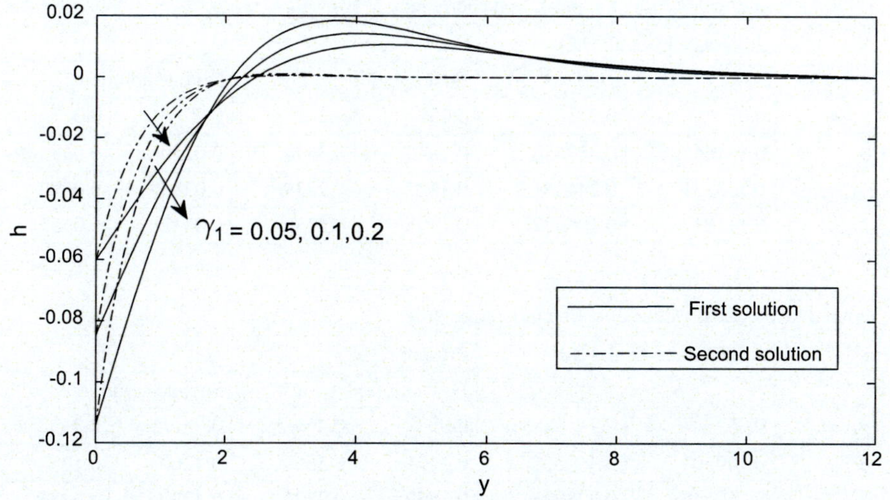

Fig. 5 Angular velocity distributions $h(y)$ for some values of γ_1 when $\Pr = 0.7$ and $K = 2$

Negative angular velocity of Newtonian fluid motion (first solution) indicates the reduction of angular displacement. Maximum variation of angular displacement is seen in the neighborhood of the surface where viscosity plays a significant role. Further, it may be concluded that magnitude of angular velocity reduces with the increase of K (Fig. 3). Physically, it may be interpreted that vortex viscosity reduces the angular momentum of governing fluid motion.

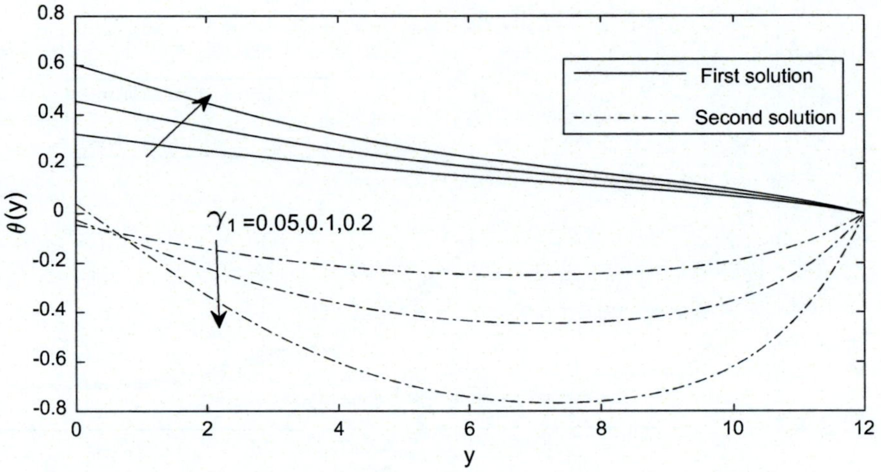

Fig. 6 Temperature distributions $\theta(y)$ for some values of γ_1 when $\mathrm{Pr} = 0.7$ and $K = 2$

Table 1 Numerical values of skin friction coefficient $f''(0)$

K↓	Alkasasbeh[a] et al. [18]		Present results			
			First solution		Second solution	
γ_1→	0.05	0.2	0.05	0.2	0.05	0.2
0	0.184661	0.357656	0.1846	0.3574	0.0150	0.0435
1	0.133231	0.244051	0.1427	0.2729	0.0283	0.0628
2	0.111617	0.195632	0.1205	0.2289	0.0312	0.0655

Table 2 Numerical values of wall temperature $\theta(0)$

K↓	Alkasasbeh[a] et al. [18]		Present results			
			First solution		Second solution	
γ_1→	0.05	0.2	0.05	0.2	0.05	0.2
0	0.149501	0.360667	0.1495	0.3606	0.4260	0.1258
1	0.157545	0.378091	0.1559	0.3726	0.4230	0.1233
2	0.162725	0.388925	0.1610	0.3820	0.0421	0.1218

Figures 4 and 5 represent the influences of the conjugate parameter γ_1 for $\mathrm{Pr} = 0.7$ and $K = 2$ on velocity and angular velocity distributions in the neighborhood of the lower stagnation point of the sphere. It is observed that fluid motion is accelerated with the conjugate parameter as the parameter (related with the heat transfer) helps to reduce the kinematic and vortex viscosities of fluid. Similarly, reduction in vortex viscosity also helps to reduce the magnitude of angular velocity of fluid.

Temperature distribution across the fluid flow is shown graphically in Fig. 6. During steady flow, there is continuous fall of temperature across the fluid flow, and the conjugate parameter enhances the temperature of fluid flow with maximum variation which is seen in the neighborhood of the surface. But, during time-dependent flow, the dimensionless temperature experiences negative values. Physically, it represents that free stream temperature is greater than temperature at the surface.

5 Conclusion

We have formulated the problem of micropolar fluid flow near the lower stagnation point with convection boundary condition. The MATLAB built-in bvp4c solver technique is used to solve the resulting problems. Some of the important results from the above investigations are highlighted below:

- When γ_1 and Pr are fixed, material parameter K helps to reduce the skin friction coefficient of steady flow (lower velocity gradient), but an opposite behavior is seen during unsteady fluid flow.
- The material parameter increases the wall temperature of steady flow, but it reduces the wall temperature during time-dependent flow.
- Fluid flow reaches its maximum speed in time-independent cases.
- Fluid flow attains its maximum variations with flow parameters in the boundary layer region.
- During unsteady flow, free stream temperature rises.
- For Newtonian fluid ($K = 0$), the steady and unsteady flow solutions of angular velocity experience completely negative values.

References

1. Eringen, A.C.: Theory of micropolar fluids. J. Math. Mech. **16**, 1–18 (1966)
2. Ariman, T., Turk, M., Sylvester, N.: Application of micro continuum fluid mechanics. Int. J. Eng. Sci. **12**, 273–293 (1974)
3. Rees, D.A.S., Bassom, A.P.: The Blasius boundary layer flow of a micropolar fluid. Int. J. Eng. Sci. **34**, 113–124 (1996)
4. Chen, T., Mucoglu, A.: Analysis of mixed force and free convection about a sphere. Int. J. Heat Mass Trans. **20**, 867–875 (1977)
5. Kasim, A.R.M., Mohammad, N.F., Aurangzaib, A., Shafie, S.: Natural convection boundary layer flow past a sphere with constant heat flux in viscoelastic fluid. J. Teknol. (2013)
6. Nazar, R., Amin, N., Grosan, T., Pop, I.: Natural convection boundary layer flow of a viscoelastic fluid on a solid sphere with Newtonian heating. Int. Commun. Heat Mass Trans **29**, 377–386 (2002)
7. Nazar, R., Amin, N., Grosan, T., Pop, I.: Free convection boundary layer on a sphere with constant surface heat flux in a micropolar fluid. Int. Commun. Heat Mass Trans. **29**, 1129–1138 (2002)

8. Cheng, C.Y.: Natural convection heat and mass transfer from a sphere in micropolarfluids with constant wall temperature and concentration. Int. Commun. Heat Mass Trans. **35**, 750–755 (2008)
9. Markin, J.: Natural-convection boundary-layer flow on a vertical surface with Newtonian heating. Int. J. Heat Fluid Flow **15**, 392–398 (1994)
10. Salleh, M.Z., Nazar, R., Pop, I.: Numerical solutions of free convection boundary layer flow on a solid sphere with Newtonian heating in a micropolar fluid. Meccanica **47**, 1261–1269 (2012)
11. Aziz, A.: A similarity solutions for laminar thermal boundary layer flow over a flat plate with a convective surface boundary condition. Commun. Nonlinear Sci. Numer. Simul. **14**, 1064–1068 (2009)
12. Makinde, O.D., Aziz, A.: MHD mixed convection from a vertical plate embedded in aporous medium with a convective boundary condition. Int. J. Therm. Sci. **49**, 1813–1820 (2010)
13. Ishak, A.: Similarity solutions for flow and heat transfer over a permeable surface with convective boundary condition. Appl. Math. Comput. **217**, 837–842 (2010)
14. Merkin, J.H., Pop, I.: The forced convection flow of a uniform stream over a flat surface with a convective surface boundary condition. Commun. Nonlinear Sci. Numer. Simul. **16**, 3602–3609 (2011)
15. Yao, S., Fang, T., Zhong, Y.: Heat transfer of a generalized stretching/shrinking wall problem with convective boundary conditions. Commun. Nonlinear Sci. Numer. Simulat. **16**, 752–760 (2011)
16. Yacob, N.A., Ishak, A.: Micropolar Fluid Flow over a Shrinking Sheet. Meccanica **47**, 293–299 (2012)
17. Mohamed, M.K.A., Salleh, M.Z., Nazar, R., Ishak, A.: Boundary Value Problems **1**, 1–10 (2013)
18. Alkasasbeh, H. T., Salleha, Mohd, Z., Tahar, R. M., Nazar, R., Pop, I.: Free convection boundary layer flow near the lower stagnation point of a solidsphere with convective boundary conditions in a micropolar fluid. AIP Conf. Proc. **1602**, 76–82 (2014)
19. SHU, J.J., Wilks, G.: Heat transfer in the flow of a cold, axisymmetric jet over a hot sphere. J. Heat Transfer-Trans. ASME **135**(3), 032201 (2013)
20. Aziz, L.A., Kasim, A. R. M., Salleh, M. Z. and Shafie, S.: Mixed convection boundary layerflow on a solid sphere in a viscoelastic micropolar fluid. In: Proceedings of the Third International Conference on Computing, Mathematics and Statistics (iCMS2017), (2019)
21. Ahmadi, G.: Self-similar solution of imcompressible micropolar boundary layer flow over a semi-infinite plate. Int. J. Eng. Sci. **14**, 639–646 (1976)
22. Weidman, P.D., Kubitschek, D.G., Davis, A.M.J.: The effects of transpiration on self-similar boundary layer flow over moving surfaces. Int. J. Eng. Sci. **44**, 730–737 (2006)
23. Khashi'ie, N.S., Arifin, N.Md., Rashidi, M.M, Hafidzuddin, E.H., Wahi, N.:Magnetohydrodynamics(MHD) stagnation point flow past a shrinking/stretching surface with double stratification effect in a porous medium. J. Thermal Ana. Calorimetry (2019)
24. Harris, S.D., Ingham, D.B., Pop, I.: Mixed convection boundary-layer flow near the stagnation point on a vertical surface in a porous medium: Brinkman model with slip. Transp. Porous Media. **77**, 267–285 (2009)

Low-Light Image Restoration Using Dehazing-Based Inverted Illumination Map Enhancement

Isha Agrawal, Teena Sharma, and Nishchal K. Verma

Abstract This paper proposes an algorithm to restore low-light images by enhancement of its inverted illumination map using dehazing. The proposed algorithm follows the Retinex theory and decomposes the low-light input image into reflectance and illumination components. The illumination map is then inverted and enhanced using dehazing-type algorithm and gamma correction. The enhanced map is again inverted back to obtain the enhanced illumination, and finally, it is combined with the original reflectance to output an enhanced image. The enhanced images obtained using the proposed algorithm have improved contrast and visual quality. The proposed method is analyzed qualitatively on various standard images used in literature and compared quantitatively with several benchmark techniques. It is found that the proposed algorithm is superior over various other benchmarks.

Keywords Low-light image restoration · Illumination map enhancement · Retinex theory · Gamma correction

1 Introduction

Images taken under low-light environment tend to exhibit a lack of contrast and poor visibility. The poor quality of such images adversely affects the efficiency of various machine vision and multimedia applications like object tracking [1] and detection [2]. As these algorithms are trained primarily on high-quality images, they yield poor

I. Agrawal (✉)
PDPM Indian Institute of Information Technology, Design and Manufacturing, Jabalpur 482005, India
e-mail: ishaagrawal2017@gmail.com

T. Sharma · N. K. Verma
Indian Institute of Technology, Kanpur 208016, India
e-mail: tee.shar6@gmail.com

N. K. Verma
e-mail: nishchal@iitk.ac.in

O. Castillo et al. (eds.), *Applied Mathematics and Computational Intelligence*, Springer Proceedings in Mathematics & Statistics 413,
https://doi.org/10.1007/978-981-19-8194-4_12

results in low-light conditions. To improve the performance of these applications, and also for aesthetic purposes, various low-light image enhancement methodologies have been proposed in recent decades.

1.1 Literature Review

One of the most intuitive approaches for low-light image enhancement and contrast restoration is the contrast enhancement-based techniques. Histogram equalization (HE) [3] constrains the image uniformly in the range of [0, 1]. However, it also enhances any noise present in the scene and gives rise to checkerboard effects. Furthermore, it fails to consider the statistical distribution of the image and, hence, leads to visible artifacts. Adaptive contrast enhancement (AHE) [4], on the other hand, is computationally expensive. Celik et al. [5] proposed a contrast enhancement method for variational and contextual information (CVC) which considers the inter-pixel relationship and their contextual information to smooth the target histogram. Lee et al. [6] used layered difference (LDR) to represent 2D histogram for contrast enhancement. They amplified the difference in gray level between neighboring pixels. Direct amplification of low-light image is another conventional approach to restore the color contrast and visibility. However, this can lead to over-enhancement and saturation of areas having bright light. It also gives rise to halos. To overcome this, Tang et al. [7] applied dehazing-type algorithm on the inverted low-light image to suppress regions with strong light from over-enhancement. Dong et al. [8] also performed dehazing on the inverted low-light images to obtain enhanced results by manipulating their illumination maps. Guo et al. introduced a restoration approach for low-light images (LIME) [9] which imposes a structure prior on the illumination map. Li et al. [10] performed super-pixel segmentation and used various statistical measures to estimate the noise level of each super-pixel. They used an adaptive approach for dehazing to avoid over-enhancement.

A benchmark model which inspired several enhancement techniques for low-light images was presented by Land et al., called as Retinex theory [11]. The theory is based on human perception and assumes that images are made up of two factors, namely reflectance and illumination. Some of the initial attempts include single [12] and multi-scale [13] Retinex-based approaches. These, however, give unnatural and over enhanced results. Shin et al. [14] introduced an algorithm to restore naturalness (ENR) which creates a mapping curve to adjust contrast and suppress edge artifacts. Fu et al. [15] used morphological operations to separate an image to its corresponding illumination and reflectance maps. They use a fusing strategy to combine the advantages of various techniques such as histogram equalization [3] and sigmoid function. Wang et al. [16] enhanced non-uniformly illuminated images by balancing naturalness and details.

Although significant advancements have been made to enhance low-light images, some of these approaches are computationally costly and, hence, unfeasible for real-time purposes. Furthermore, although Retinex theory preserves the edge information and the naturalness of an image, the existing algorithms still fail to completely restore an image without any loss of contrast or color information.

1.2 Contributions

While several methodologies have been developed to restore low-light images, they often tend to lose color information or do not perform significant enhancement or improvements. This paper, on the other hand, proposes a simple, yet efficient approach to effectively enhance and restore a given low-light image. The dehazing-type approach used in the paper increases the general pixel intensity of the image. Also, using gamma correction further widens the dynamic range of the image, thereby improving its contrast and overall visual appeal. The results obtained present the efficiency of the proposed method when compared with several benchmark techniques in qualitative as well as quantitative terms.

The remaining of the paper is arranged as follows: Sect. 2 illustrates upon the algorithm introduced and describes it in detail. Section 3 includes the qualitative and quantitative results obtained and their comparison with several benchmarks. Lastly, Sect. 4 highlights some of the important discussions followed by the concluding remarks.

2 Proposed Methodology

This section includes the explanation of the proposed methodology. Figure 1 outlines the steps to be followed in the proposed algorithm. The following subsections involve discussions on the basic building blocks of Fig. 1.

2.1 Retinex Theory

The basic assumption of the Retinex theory [11] is that images can be separated into their constituent illumination (lightness) and reflectance maps. Mathematically, an input image I_{input} can be represented by its illumination L_{illum} and reflectance R_{ref} components as

$$I^c_{\text{input}}(a, b) = R^c_{\text{ref}}(a, b) \times L_{\text{illum}}(a, b) \tag{1}$$

where (a, b) is the pixel location and c is for the blue (B), green (G), and red (R) color channels. It can be observed that the reflectance is different for different color

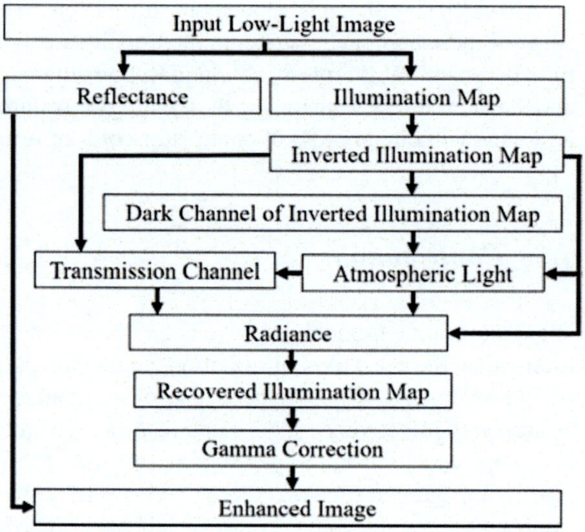

Fig. 1 Block diagram of the proposed methodology

channels as it represents the properties of each color. On the other hand, the illumination map for all the color channels is the same, as the same amount of light reaching each of the color channels. From (1), the reflectance is defined as

$$R^c_{\text{ref}}(a,b) = \frac{I^c_{\text{input}}(a,b)}{L_{\text{illum}}(a,b)}. \tag{2}$$

2.2 *Illumination Map Estimation*

The illumination map can be approximated as the maximum intensity of the image at each pixel location. For an input image I_{input}, the initial illumination map L_{illum} is estimated as

$$L_{\text{illum}}(a,b) = \max_{c\in\{B,G,R\}} I^c_{\text{input}}(a,b) \tag{3}$$

The inverted illumination L_{inv} can then be calculated as

$$L_{\text{inv}}(a,b) = 255 - L_{\text{illum}}(a,b) \tag{4}$$

Herein, it has been assumed that the range of the input image is [0, 255]. Figures 2a–c show a low-light image, its illumination map using (3), and the inverted illumination map using (4), respectively. It can be observed that the high intensity of the inverted illumination map gives it a hazy appearance. Accordingly, suitable dehazing techniques can be applied to the inverted illumination map to enhance it.

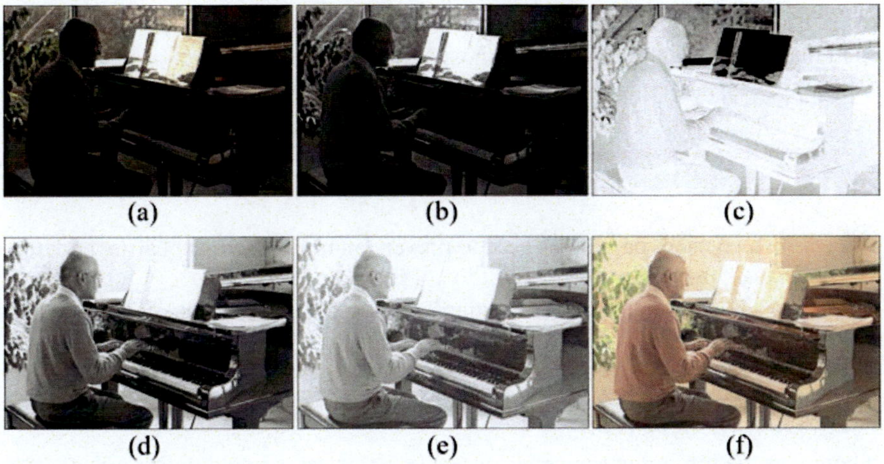

Fig. 2 Image enhancement using proposed algorithm: **a** input image, **b** corresponding illumination map approximated by (3), **c** inverted illumination map calculated by (4), **d** recovered illumination map using (9), **e** illumination map further enhanced with gamma correction by (10), and **f** final enhanced output image by (11)

2.3 Dehazing of Inverted Illumination Map

As the inverted illumination map has a hazy appearance, dehazing like approach can be applied to it. A pioneer dehazing technique is based on the conventional atmospheric scattering model [17], given by

$$I^c(a,b) = t_{\text{trans}}(a,b) \times J^c(a,b) + A^c(1 - t_{\text{trans}}(a,b)) \tag{5}$$

where $J(a,b)$ is the original dehazed scene, $I(a,b)$ represents the observed image, $t_{\text{trans}}(a,b)$ represents the transmission map representing the amount of light that reaches the observer, A is global atmospheric light, a and b represent the coordinate locations of the pixels, and c represents the color channels B, G, and R. The global atmospheric light represents the pixels with maximum light in an image. To calculate the atmospheric light, the brightest 0.1% pixels of the dark channel [18] are considered. For an inverted illumination map L_{inv}, the corresponding dark channel [18] $L_{\text{inv}}^{\text{dark}}$ is given by

$$L_{\text{inv}}^{\text{dark}}(a,b) = \min_{(a,b)\in\Omega(i,j)} (L_{\text{inv}}(a,b)) \tag{6}$$

where Ω represents a patch with center at pixel (i, j). The patch size considered here is 3×3.

The pixel locations corresponding to the brightest pixel intensities in the dark channel are mapped to their corresponding locations of the pixels in the illumination map. The highest intensity among these, i.e., 0.1% pixel locations in the inverted illumination map is the estimated atmospheric light.

The transmission map can then be approximated using atmospheric light as

$$t_{\text{trans}}(a, b) = 1 - \omega \left(\frac{L_{\text{inv}}^{\text{dark}}(a, b)}{A} \right) \tag{7}$$

where ω is a constant parameter used to prevent an image from appearing unnatural. Based on [18], ω is set to 0.95.

Finally, the dehazed inverted illumination map $L_{\text{inv}}^{\text{dehazed}}$ is calculated as

$$L_{\text{inv}}^{\text{dehazed}}(a, b) = A + \frac{(L_{\text{inv}}(a, b) - A)}{\max(t_{\text{trans}}(a, b), t_0)}, \quad t_0 = 0.1 \tag{8}$$

where t_0 is the minimum transmission value that is set to avoid division by 0.

The enhanced illumination map L_{enh} of the input low-light image is then obtained as

$$L_{\text{enh}}(a, b) = 255 - L_{\text{inv}}^{\text{dehazed}}(a, b) \tag{9}$$

This illumination map can further be enhanced to improve the contrast by applying gamma correction as

$$L_{\text{enh}}^{\text{gamma}}(a, b) = (L_{\text{enh}}(a, b))^{\gamma} \tag{10}$$

where $\gamma = 0.5$ is the gamma correction parameter.

Finally, the required enhanced output image I_{enh} is calculated as

$$I_{\text{enh}}^{c}(a, b) = R_{\text{ref}}^{c}(a, b) \times L_{\text{enh}}^{\text{gamma}}(a, b) \tag{11}$$

where $R_{\text{ref}}^{c}(a, b)$ is the reflectance of the image calculated using (2).

3 Experimental Results

The proposed algorithm has been compared with several state of the art[1], namely HE [3], AHE [4], CVC [5], LDR [6], ENR [14], and LIME [9]. The results[2] suggest the superiority of the proposed methodology over the considered benchmarks both quantitatively and qualitatively.

---

[1] The results of the state of the art have been obtained using MATLAB libraries or the source codes made publicly available by the authors.

[2] The results of the proposed algorithm on images in Fig. 6 are available at https://drive.google.com/drive/u/0/folders/1jf_uq_ua5JpPmxJtQAyQzL1WDntue7ct.

3.1 Qualitative Analysis

Figures 3, 4, and 5 present the output of the proposed methodology along with several benchmarks. It can be observed that while the result of the proposed approach has sufficient light and color information, AHE [4], CVC [5], and LDR [6] fail to preserve any such color fidelity. Although the result of HE [3] in Fig. 3b has increased overall pixel intensity, the image appears whitewashed. Outputs of CVC [5] and LDR [6] continue to appear dark and fail to enhance the contrast of the image, as can be observed in images (d) and (e) of Figs. 3, 4, and 5. The proposed algorithm performs better than these benchmarks in terms of visual appeal while giving comparable results as ENR [14] and LIME [9].

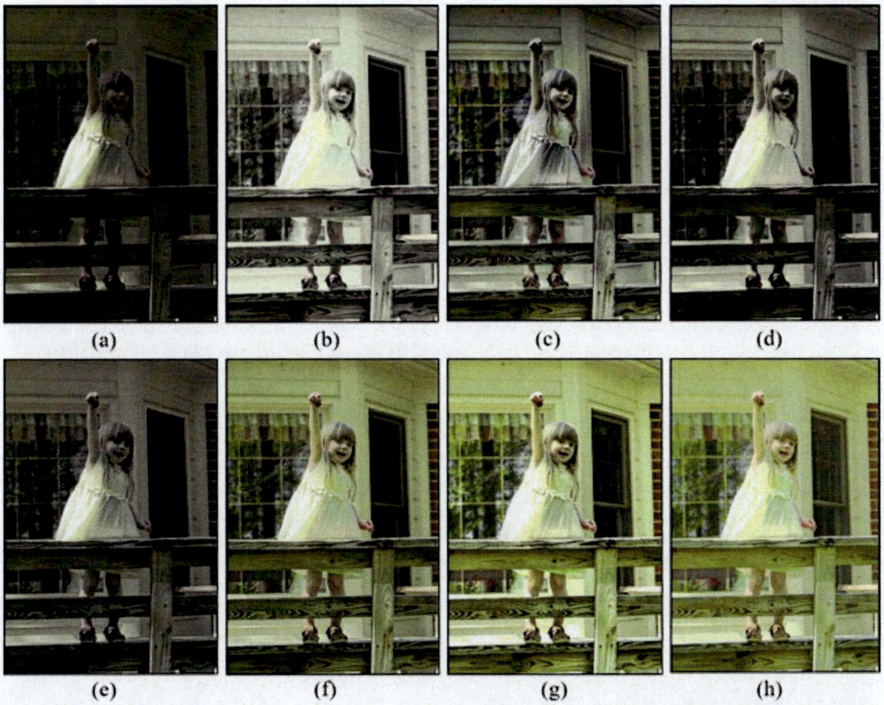

Fig. 3 Comparison of proposed methodology with various benchmarks on Waving Girl image: **a** input low-light image, outputs of **b** HE [3], **c** AHE [4], **d** CVC [5], **e** LDR [6], **f** ENR [14], **g** LIME [9], and **h** proposed method

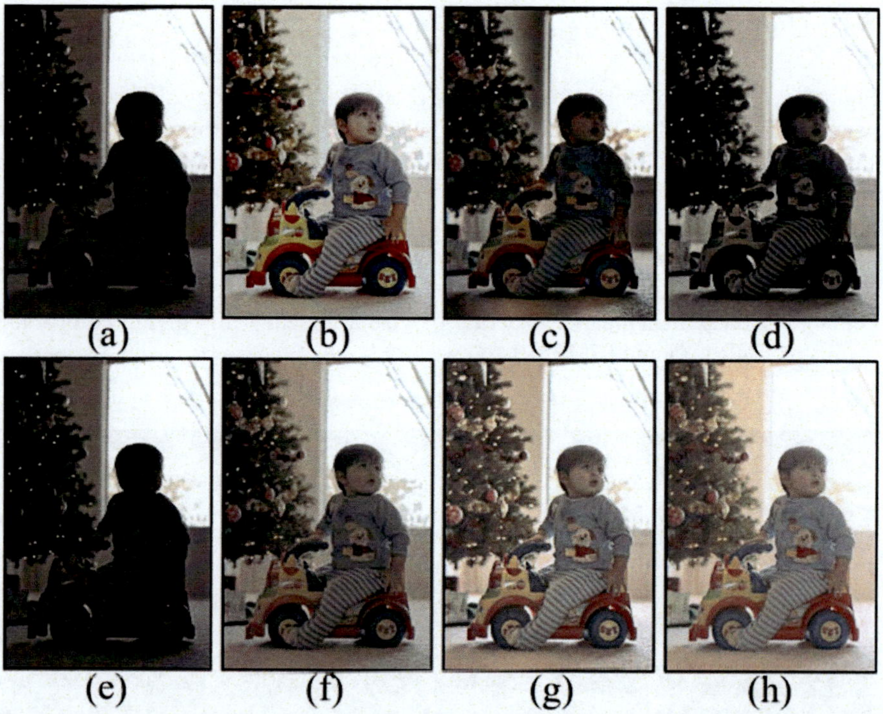

Fig. 4 Comparison of proposed methodology with various benchmarks on Christmas Rider image: **a** input low-light image, outputs of **b** HE [3], **c** AHE [4], **d** CVC [5], **e** LDR [6], **f** ENR [14], **g** LIME [9], and **h** proposed method

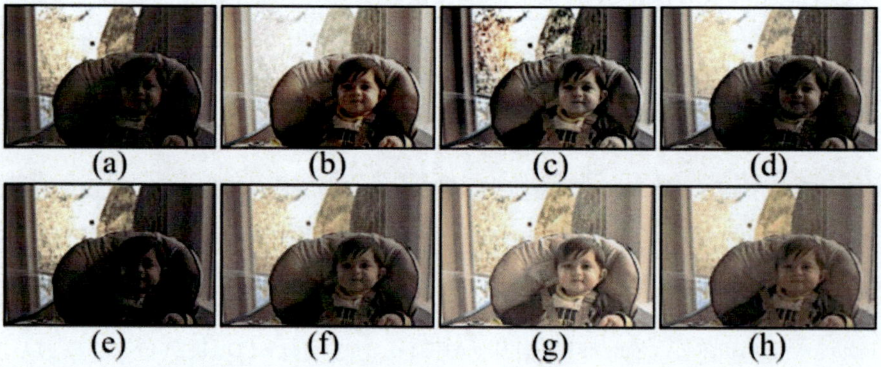

Fig. 5 Comparison of proposed methodology with various benchmarks on High Chair image: **a** input low-light image, outputs of **b** HE [3], **c** AHE [4], **d** CVC [5], **e** LDR [6], **f** ENR [14], **g** LIME [9], and **h** proposed method

Fig. 6 Various standard images used for quantitative comparison. **From left to right and top to bottom:** Images of Baby at Window, Baby on Grass, Christmas Rider, High Chair, Standing Boy, Waving Girl, Dog, and Santa's Little Helper. (Download Link: Click here)

Table 1 Performance comparison in terms of LOE ($\times$ 10^3) [16]

Image name	HE [3]	AHE [4]	CVC [5]	LDR [6]	ENR [14]	LIME [9]	Proposed
Baby at Window	0.128	2.234	0.164	0.158	**0.073**	1.473	0.082
Baby on Glass	0.146	2.902	0.246	0.132	0.109	2.016	**0.055**
Christmas Rider	0.247	1.297	0.446	0.253	0.224	0.952	**0.144**
High Chair	0.122	2.471	0.191	0.110	0.804	1.580	**0.063**
Standing Boy	0.395	2.038	0.157	0.141	0.102	1.348	**0.084**
Waving Girl	0.222	0.186	0.182	0.982	0.116	0.994	**0.095**
Dog	0.235	1.304	0.229	0.110	0.056	1.132	**0.015**
Santa's Little Helper	0.191	1.390	0.294	0.184	**0.144**	1.391	0.156
Average	0.211	1.728	0.239	0.259	0.204	1.361	**0.087**

Numbers in bold represent the best performance amongst all considered methods for that image

3.2 Quantitative Analysis

The proposed method has been compared with several benchmarks on various low-light images shown in Fig. 6. The performance metric used for comparison is the lightness order error (LOE) [16], mathematically given by

$$\text{LOE} = \frac{1}{m}\sum_{a=1}^{m}\sum_{b=1}^{m}\left(U\left(M_{\mathrm{e}}(a), M_{\mathrm{e}}(b)\right) \oplus U\left(M_{\mathrm{r}}(a), M_{\mathrm{r}}(b)\right)\right) \tag{12}$$

where m is the pixel location, $U(\alpha, \beta) = 1$ if $\alpha > \beta$, and 0 if $\alpha \leq \beta$, $\oplus$ represents the exclusive OR operation, and $M_r(i)$ and $M_e(i)$ are the highest pixel values among blue, green, and red color channels of the reference and enhanced images, respectively, at pixel location i. Lower value of LOE [16] indicates higher naturalness of lightness in the images.

It can be inferred from Table 1 that the proposed method performs better than nearly all the considered benchmarks. Although ENR [14] outperforms the proposed algorithm in two images (Baby at Window and Santa's Little Helper), the proposed algorithm gives superior results in most of the cases. In the image of High Chair, the LOE value of the proposed method is lesser than one-tenth of the LOE value of ENR [14] and is lower by a factor of nearly two in the image of Baby on Grass and Christmas Rider. The proposed method gives better results than nearly all the benchmarks considered in qualitative as well as quantitative parameters.

4 Conclusions

The paper proposed an image restoration approach for low-light images by enhancing its illumination map. The results show the superiority of the proposed methodology over various other famous benchmarks. It can be concluded that the proposed method gives visually pleasing results with enhanced contrast and color fidelity. The qualitative and quantitative results are superior to a majority of the other benchmark methods considered. However, the future work may include estimation of the enhanced color image directly with input low-light image.

References

1. Zhang, K., Zhang, L., Yang, M.: Real-time compressive tracking. In: Proceedings of the $12^t h$ European conference on Computer Vision, pp. 864–877, Florence, Italy, 7–13 Oct (2012). https://doi.org/10.1007/978-3-642-33712-3_62
2. Oneata, D., Revaud, J., Verbeek, J., Schmid, C.: Spatio-temporal object detection proposals. In: ECCV-European Conference on Computer Vision, pp. 737–752, Zurich, Switzerland, Sep (2014). https://doi.org/10.1007/978-3-319-10578-9_48

3. Gonzalez, R.C., Woods, R.E.: Digital Image Processing, 3rd edn. Prentice-Hall, Englewood Cliffs, NJ, USA (2007)
4. Zuiderveld, K.: Contrast limited adaptive histogram equalization. In: Graphic Gems IV, pp. 474–485. Academic Press Professional, San Diego, Aug (1994)
5. Celik, T., Tjahjadi, T.: Contextual and variational contrast enhancement. IEEE Trans. Image Proc. **20**(12), 3431–3441, Dec (2011). https://doi.org/10.1109/TIP.2011.2157513
6. Lee, C., Lee, C., Kim, C.-S.: Contrast enhancement based on layered difference representation of 2D histograms. IEEE Trans. Image Process 22(12), pp. 5372–5384, Dec 2013. https://doi.org/10.1109/TIP.2013.2284059
7. Tang, C., Wang, Y., Feng, H., Xu, Z., Li, Q., Chen, Y.: Low-light image enhancement with strong light weakening and bright halo suppressing. IET **13**(3), 537–542, 28 Feb (2019). https://doi.org/10.1049/iet-ipr.2018.5505
8. Dong, X., Wang, G., Pang, Y., Li, W., Wen, J., Meng, W., Lu, Y.: Fast efficient algorithm for enhancement of low lighting video. In: 2011 IEEE International Conference on Multimedia and Expo. Barcelona, Spain, 11–15 July (2011). https://doi.org/10.1109/ICME.2011.6012107
9. Guo, X., Li, Y., Ling, H.: LIME: Low-light image enhancement via illumination map estimation. IEEE Trans. Image Process. **26**(2), 982–993, Feb (2017). https://doi.org/10.1109/TIP.2016.2639450
10. Li, L., Wang, R., Wang, W., Gao, W.: A low-light image enhancement method for both denoising and contrast enlarging. In: 2015 IEEE International Conference on Image Processing (ICIP). Quebec City, QC, Canada, 27–30 Sep 2015. https://doi.org/10.1109/ICIP.2015.7351501
11. Land, E.H., McCann, J.J.: Lightness and retinex theory. J. Opt. Soc. Am. **61**(1), 1–11, (1971). https://doi.org/10.1364/JOSA.61.000001
12. Jobson, D.J., Rahman, Z., Woodell, G.A.: Properties and performance of acenter/surround retinex. IEEE Trans. Image Proc. **6**(3), 451–462, Mar (1997). https://doi.org/10.1109/83.557356
13. Jobson, D.J., Rahman, Z., Woodell, G.A.: A multiscale Retinex for bridging the gap between color images and the human observation of scenes. IEEE Trans. Image Proc. **6**(7), 965–976, July (1997). https://doi.org/10.1109/83.597272
14. Shin, Y., Jeong, S., Lee, S.: Efficient naturalness restoration for non-uniform illumination images. IET Image Proc. **9**(8), 662–671, 27 July (2015). https://doi.org/10.1049/iet-ipr.2014.0437
15. Fu, X., Zeng, D., Huang, Y., Liao. Y., Ding, X., Paisley, J.: A fusion-based enhancing method for weakly illuminated images. Signal Process. **129**, 82–96, Dec (2016). https://doi.org/10.1016/.sigpro.2016.05.031
16. Wang, S., Zheng. J., Hu, H.-M., Li, B.: Naturalness preserved enhancement algorithm for non-uniform illumination images. IEEE Trans. Image Process. **22**(9), 3538–3548, Sep (2013). https://doi.org/10.1109/TIP.2013.2261309
17. Narasimhan, S.G., Nayar, S.K.: Vision and the atmosphere. Int. J. Comput. Vis. **48**(3), 233–254 (2002). https://doi.org/10.1023/A:1016328200723
18. He, K., Sun, J., Tang, X.: Single image haze removal using dark channel prior. IEEE Trans. Pattern Anal. Mach. Intell. **33**(12), 2341–2353, Dec (2011). https://doi.org/10.1109/TPAMI.2010.168

Mathematical Modeling of Probability and Profit of Single-Zero Roulette to Enhance Understanding of Bets

Arjun Venkatesh and R. Gomathi Bhavani

Abstract Roulette is one of the very popular casino games, and the game presents itself as an excellent candidate for demonstrating the application of mathematical analysis to explore probabilistic reasoning and profitability of bets placed. This work takes up the case of single-zero roulette to form a mathematical model for understanding the winning bet. Firstly, the bets are put into forms of probabilistic events for building the model. Then grouping these into sets, several results from set theory are made use of to generalize bets. The results are then observed from the perspective of understanding the profit, given the outcome of a spin. The modeling for simple bet just gave a generalization and did not increase the probability of winning or not losing the game. Modeling complex bets, on the other hand, turned out to be extremely useful to enhance understanding of bets. The results showed that the probability increased with the change in bet amount which meant that the modeling enhanced players' understanding of bets, though it did not suggest a winning strategy. This enables the player to have a better decision and choice of the bets he or she wants to place, beforehand.

Keywords Bets · Gambling · Probability · Profit · Roulette

1 Introduction

People's lives in the modern-day work culture are scheduled to the maximum, and hence, some of them seek leisure and entertainment as a means to deviate from the routine and also to alleviate stress and boredom. Some of the leisurely activities involve thinking and thus enhance the neural patterns and wiring, while some relax the brain. The casino games fall under the category wherein people play them due to their need for speed, adventure, skill, and pastime. Overcoming loneliness, earning quick money, need for some social interaction and relaxation, wanting some fun, and falling prey to addiction are said to be some of the reasons why people resort

A. Venkatesh · R. Gomathi Bhavani (✉)
Department of Electrical and Electronics Engineering, Birla Institute of Technology and Science Pilani, Dubai Campus, Dubai, UAE

O. Castillo et al. (eds.), *Applied Mathematics and Computational Intelligence*, Springer Proceedings in Mathematics & Statistics 413,
https://doi.org/10.1007/978-981-19-8194-4_13

to playing casino games. These reasons might differ along the lines of gender and demography. Sensation seeking and superstitious thinking were consistent predictors across gender, while probabilistic reasoning ability, the perception of the economic profitability of gambling, and peer gambling behavior were found to be predictors only among male adolescents, whereas parental gambling behavior had a predictive power in female adolescents [1]. Roulette is one of the very popular casino games, the rules of which vary slightly according to the geographic location in which it is played. In the Europe, as well as Australia, roulette is played with 37 numbered slots consisting of eighteen red slots, eighteen black slots, and one green slot. Several literatures [2–4] presented statistical analysis and observed that winning for an extended run is a matter of less probability. Although none of the literature has come up with sure-shot strategies to help eliminate the uncertainty associated with playing the game to a favorable outcome, the game presents itself as an excellent candidate to demonstrate the application of mathematical formulae and analysis.

2 Related Works

A review of the literature reveals the existence of a limited number of studies done on modeling of roulette bets. Barboianu [5] presented a collection of odds and figures attached to a large range of complex bets, revealed in their entire mathematical structure. This book provided just mathematical facts and not so-called winning strategies. Croucher [6] discussed some strategies for playing roulette, making use of binomial distribution and normal approximation. Roulette, as with most games of chance, also provides an excellent teaching opportunity in demonstrating an interesting and practical use of the normal approximation in the calculation of binomial probabilities [6]. Croson and Sundali [7] presented results from the field, using videotapes of patrons gambling in a casino, to examine the existence and extent of the biases, namely the gambler's fallacy and the hot hand, in naturalistic settings. The paper presented datasets in the form of descriptive statistics of the wheel and the bets placed. Small and Tse [8] provided a short review of the history and then set out to determine to what extent that determinism can really be exploited for profit. With a more sophisticated, albeit more intrusive, system (mounting a digital camera above the wheel), the paper demonstrated a range of systematic and statistically significant biases which could be exploited to provide an improved guess of the outcome. Lichtenstein et al. [9] examined the effect of instruction in expected value on optimality of gambling decisions and reported that the expected value is to be irrelevant as a guide to action, even when the concept was carefully explained and clearly displayed for them. Abuhanna [10] generated data using the physical roulette wheel to test whether an association exists between initial conditions (the pocket from which the ball is released) and output (the pocket where the ball lands). Data were generated to determine whether there exists statistical significance in distributions of adjoined pockets [10]. Pflaumer [11] presented important formulas for the martingale strategy, such as the distribution, the expected value, the standard deviation of the profit, the risk of a loss, or the expected bet of one or multiple martingale rounds. The work examined a martingale or doubling strategy as a way of improving the chances of winning a roulette game. Seal and Przasnyski [12] implemented a model of roulette in a spreadsheet that can simulate outcomes of various betting strategies and used the model to simulate the Martingale strategy. Kavokin et al. [13] estimated the critical starting capital needed to ensure the low-risk game for an indefinite time and showed that the dramatic increase of gambler's chances was a manifestation of bunching of numbers in a non-ideal roulette. The next section describes systematic modeling procedure that was developed for this work.

3 Mathematical Modeling

Roulette is one of the very popular casino games, the rules of which vary slightly from country to country. In the Europe, as well as Australia, roulette is played with 37 numbered slots consisting of eighteen red slots, eighteen black slots, and one green slot. The green slot has a single zero marked, while in the USA, two green slots are used, that are numbered with a double as well as a single zero. Roulette, unlike all other casino games, does not involve tricks like guessing cards or a dealer cheating with cards, and neither does it involve electronic slot machines, rigged to make a player lose. This work takes up the case of single-zero roulette to form a mathematical model for understanding the winning bet. Firstly, the bets are put into forms of probabilistic events for building the model. Then grouping these into sets, many results from set theory are made use of to generalize bets. The results are then observed from the perspective of understanding the profit, given the outcome of a spin. In the upcoming subsection, a glance of rules of European single-zero roulette and the kinds of bets allowed are presented.

3.1 Simple Bets

Roulette is played with a wheel and a betting layout. A dealer is at a table, and a spinning wheel, numbered 0–36, is rolled on the table as shown in Fig. 1. Then, a ball is thrown along the edge of the spinning wheel. Bets are placed on a particular number or color the ball would land on. The winner is declared publicly, if the ball lands on any number included in the set of numbers bet on. A fresh round of betting begins after clearing away the lost ones.

Although there are many types of bets that are allowed, the following summarize the most popular ones: red/black (where one decides to bet on red or black numbers), high/low (where the number is in 1–18 or 19–35), odd/even (bet on odd numbers or even numbers), columns (where one bets on column 1 of 1, 4, 7, 10, 13, 16, 19, 22, 25, 28, 31, 34 or column 2 or column 3), dozens (this bet is between first 12, second 12, or third 12), straight up (this bet is placed on one number), split (this bet is placed

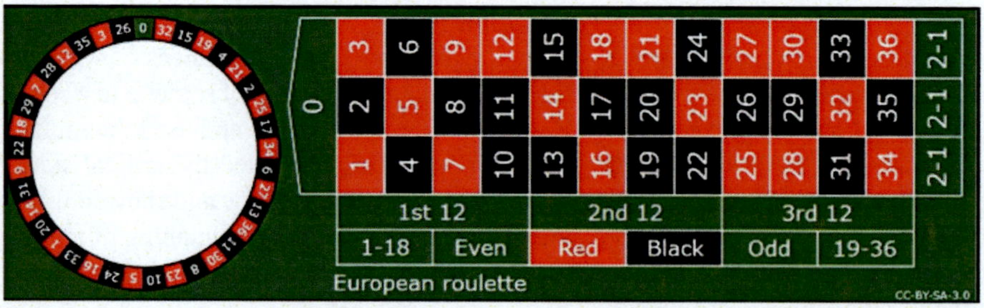

Fig. 1 Wheel and the corresponding layout

on a line joining two numbers; if either number comes up, one wins), street (this bet is placed on a row consisting of three numbers such as 123, 456), corner (this bet is similar to split bet, bet on four numbers, and the ball landing on any one of those numbers means winning bet), and line (these are like street bets, but the bet is on two rows, and hence, the ball landing on one of the six numbers constitutes a winning bet). All of these bets will be a losing one if the ball lands on the green zero. Since there are 37 equally likely possibilities, it is easy to see that the probability that any one of the above bets is successful on a single spin is 18/37 or 0.486486. So, firstly, to build a mathematical model for the profit, all possible elements are added to a sample space and made this a set, with R—referring to all possible roulette numbers of European roulette.

$$R = \{r | r \in Z^{+}, 0 \leq r \leq 36\} \tag{1}$$

A simple bet can be defined as a bet that is made by a unique placement of a chip. A simple bet means that the chips can only be placed for one bet. When one places bets in roulette, generally the main interest point comes up in winning or losing. This does in fact depend not only on impacts of mathematics and probability but also on the player's style and keenness in profit. Generally, in every European casino roulette, a split bet on {0,1}{0,2}{0,3} is allowed, and there are 12×2 horizontal placements and 11×3 vertical placements. In Fig. 2, different colors indicate the different kinds of placements on the layout. For example, a street bet is placed on (13, 14, 15). This has a probability of winning of 3/37, and if the bet is won, one is given eleven times the bet amount, plus the initial bet. Denoting all the roulette numbers by R, some generalizations are made. Hence, any allowed bet must be a subset of R. Let us denote the group of sets of all the allowed bets, and that a simple bet with one placement as B. As an example, straight-up bet is modeled as $\{5\} \in B$ and split bet as $\{8, 5\} \in B$.

Even bets (note zero is not counted)

$$\{2, 4, 6, 8, 10, 12, 14, 16, 18, 20, 22, 24, 26, 28, 30, 32, 34, 36\} \in B \tag{2}$$

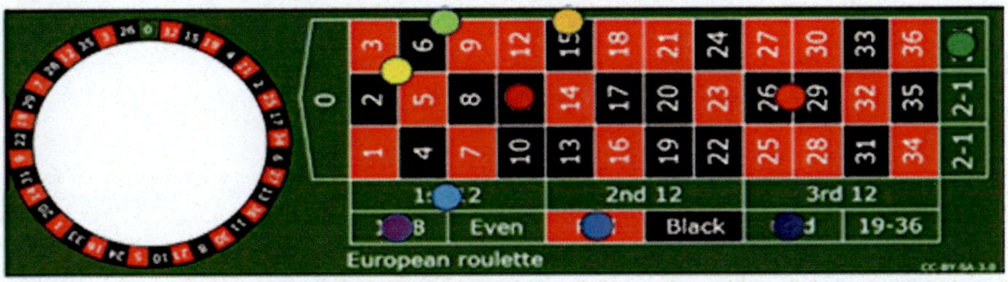

Fig. 2 Placement of various bets in the layout

However, the following sets cannot be made:

$$\{5, 26, 15\} \notin B \tag{3}$$

$$\{2, 35\} \notin B \tag{4}$$

Hence, a simple bet is put in the form of a triplet of numbers, denoted as (b, W_C, A), where W_C is the winning coefficient, or the number one's bet gets multiplied by if one wins, also called as the payout, $b \in B$ b is the bet placement and A is the amount bet in chips. Quite clearly, for European single-zero game, the $W_C \in \{1, 2, 5, 8, 11, 17, 35\}$ as these are the payouts.

Hence, the probability of winning a simple bet can be modeled as:

$$P(b) = \frac{\#b}{\#R} \tag{5}$$

Here, *#b* is the cardinality of the set b and depends on what bet b is placed by the person. Here, $\#R = 37$, as there are 37 possible numbers. Once the spin has been made, there are two possible situations: either player wins the bet b and gets a profit of $W_C A$ or the player loses bet b if the number is from R-b and loses A. For a simple bet, the describing equation is written as:

$$P_{bet} = f(x) W_C A - (1 - f(x)) A \tag{6}$$

Here, P_{bet} is the profit of the bet of amount A made, and $f(x)$ is a characteristic function that depends on bet b and the outcome of the spin x. $f(x)$ is defined as follows:

$$f(x) = \begin{cases} 1, if x \in b \\ 0, if x \notin b \end{cases} \tag{7}$$

Below, an example of a couple of simple bets is taken, and the probability and profit of the bet are calculated. Assuming a player makes a street bet of the row shown with numbers 16,17,18 with reference to Fig. 2 with an amount of 5$,

$$b = \{16, 17, 18\},\ W_C = 11,\ A = 5\$ \tag{8}$$

The probability of winning this bet can be calculated as:

$$P(b) = \frac{\#b}{\#R} = \frac{3}{37} = 0.081 = 8.1\% \tag{9}$$

The profit of the bet will be

$$P_{\text{bet}} = f(x)W_C A - (1 - f(x))A \tag{10}$$

$$P_{\text{bet}} = \begin{cases} 50 & \text{if } x \in \{16, 17, 18\} \\ -5 & \text{if } x \notin \{16, 17, 18\} \end{cases} \tag{11}$$

The probability and payout thus calculated are presented in Table 1.
The expected profit of a simple bet can be calculated as:

$$E(P_{\text{bet}}) = W_C{*}A{*}P(b) - A(1 - P(b)) \tag{12}$$

where $E(P_{\text{bet}})$ is the expected profit of a simple bet. A is the amount, and $P(b)$ is the probability of winning bet b. For example, a bet is placed on line bet

$$b = \{1, 2, 3, 4, 5, 6\},\ A = 3\$,\ W_C = 5 \tag{13}$$

For this, $P(b) = {}^{6}/_{37}$. Hence, the expectation, $E(P_{bet})$ = -0.082.

Table 2 presents the expectation of simple bets for amounts 2 to 10$, in the multiples of 2. It is clear that the modeling of simple bets did not yield any significant information as the expectations of all bets are equal. However, it can be observed that all values are negative. This is because of house edge, often found in casinos.

Table 1 Probabilities and payouts of simple bets

Serial no	Simple bet	Probability	Payout	Possible number of placements
1	Straight up	1/37 = 2.70%(36:1)	35–1	37
2	Split	2/37 = 5.40%(17.5:1)	17–1	60
3	Street	3/37 = 8.10%(11.3:1)	11–1	12 (12 rows)
4	Corner	4/37 = 10.81%(8.2:1)	8–1	22 (11 × 2 corners)
5	Line	6/37 = 16.21%(5.1:1)	5–1	11 (11 nodes with number)
6	Column	12/37 = 32.43%(2:1)	2–1	3 (1 per column)
7	Dozen	12/37 = 32.43%(2:1)	2–1	3 (1 per dozen)
8	Color	18/37 = 48.64%(1:1)	1–1	2 (1 per color)
9	Even/odd	18/37 = 48.64%(1:1)	1–1	2 (even or odd)
10	Low/high	18/37 = 48.64%(1:1)	1–1	2 (first 18 or second 18)

Table 2 Expected value of profits of simple bets

Simple bet	Probability (%)	Payout	Amount (US$)				
			2	4	6	8	10
Straight up	2.70	35	−0.054	−0.108	−0.162	−0.216	0.270
Line	16.21	5	−0.054	−0.108	−0.162	−0.216	0.270
Low/high	48.65	1	−0.054	−0.108	−0.162	−0.216	0.270

Ideally, we would like to have the profit of a bet as reciprocal of the probability. But, people often ignore the zero, which is casino's edge.

A simple bet means only one bet can be made; however, in roulette's rules, a player can make many bets at the same time in the same spin. These complex bets can also be called simultaneous placement of simple bets. Enlisting each of the complex bets would be extremely tedious to do, because the set B has 154 elements; hence, the power set of B will have 2154 elements, and this number is a very large number with about 47 digits. Hence, we denote each complex bet with a finite group of triplets.

3.2 Complex Bets

For modeling complex bets, a finite group is defined as given under:

$$I = \left(b_y, A_y, W_{Cy}\right)_{y \in Y} \tag{14}$$

Here, $b_y \epsilon B$ and b_y is the bet placed. A_y is the amount of chips for the respective bet, and W_{Cy} is the winning coefficient or the profit for each bet, and this is defined for every $y \in Y$. Y is a finite set of consecutive indexes, consisting of a set, starting from 1 till the number of simple bets made. If $\#Y = 1$, then the bet is simple. For a complex bet I as defined above, we assume that the outcome of x belongs to sets $b_1, b_2, b_3, \ldots, b_m$, but these complex bets do not belong to the sets, $b_{m+1}, b_{m+2}, b_{m+3}, \ldots, b_n$. Hence, in this case, the profit of the player can be described by the equation:

$$W_{C1}A_1 + W_{C2}A_2 + W_{C3}A_3 + \ldots W_{Cm}A_m - A_{m+1} - \ldots A_n \tag{15}$$

Here again, the profit can be negative. Going by the same procedure as was done for simple bets, the profit is calculated as follows: For a complex bet $I = \left(b_y, A_y\right)_{y \in Y}$, the profit is found out by adding all the profits of every individual bets.

$$P_{\text{Cbet}} = \sum_{y \in Y} \left[f_y(x) W_{Cy} A_y - \left(1 - f_y(x)\right) A_y\right] \tag{16}$$

where P_{Cbet} is the profit of complex bet, A_y is the amount of the bets, W_{Cy} is the winning coefficient for each of the y bets, and $f_y(x)$ is a characteristic function, for each of the y numbers, such that

$$f_y(x) = \begin{cases} 1, & \text{if} x \in b \\ 0, & \text{if} x \notin b \end{cases} \tag{17}$$

It is worth recalling the fact that for every one of the simple bets in the complex bet, *f*(*x*) can be 1 or 0. For each of the bets, the profit can be negative or positive. In addition, it is worth observing that the following four cases can arise:

Case 1: If the bet $I = \left(b_y, A_y\right)_{y\in Y}$ is disjoint, then P_{Cbet} is constant for each b_y. This is quite natural, since sets b_y are all mutually exclusive on each other, and the outcome of the spin $x \in b_y$ belongs only to b_y and does not belong to any of the sets b_z, $(z \neq y)$. Then, for every $x \in b_y$,

$$P_{Cbet} = A_y W_{Cy} - \sum_{z\in(Y-\{y\})} A_z \tag{18}$$

Case 2: P_{Cbet} is constant on $R - \bigcup_{y\in Y} b_y$

This is for any family of sets from *B*. By intuition, players tend to believe that the higher the numbers, the better is the profit, but here, it is shown that the profit depends on everything other than the numbers covered.

$$\text{If } x \in R - \bigcup_{y\in Y} b_y, \text{ then } x \notin b_y \text{ for any } y \in Y \tag{19}$$

Hence, the equation for profit can be written as

$$P_{Cbet} = -\sum_{y\in Y} A_y \tag{20}$$

This value is negative; hence, the value is always a loss when the spin outcome (x) is not included in the placement b_y. While writing the above equations, it was assumed that each of the sets b_y is mutually exclusive. But realistically speaking, while making complex bets, a player does not tend to make these bets mutually exclusive. Hence to find the probability of their unions accurately to mimic realistic situations, the inclusion–exclusion principle is applied by taking all the intersections between these simple bets, which makes it complex.

The sets BL, LO, and $3C$ are defined to indicate black, low, and third column, respectively. Referring to Fig. 1,

$$\text{BL} = \{2, 4, 6, 8, 10, 11, 13, 15, 17, 19, 20, 22, 24, 26, 29, 31, 33, 35\} \tag{21}$$

$$\text{LO} = \{1, 2, 3, 4, 5, 6, 7, 8, 9, 10, 11, 12, 13, 14, 15, 16, 17, 18\} \quad (22)$$

$$3\text{C} = \{3, 6, 9, 12, 15, 18, 21, 24, 27, 30, 33, 36\} \quad (23)$$

Using the inclusion–exclusion principle,

$$\begin{aligned} p(BL \cup LO \cup 3C) &= p(BL) + p(LO) + p(3C) - p(BL \cap LO) \\ &\quad - p(LO \cap 3C) - p(BL \cap 3C) + p(BL \cap LO \cap 3C) = 83.8\% \end{aligned} \quad (24)$$

This probability does not mean that a player would win the placed bet 83.8% of the time; it only indicates that the spin outcome covers one of the numbers bet on. There is still a possibility that the profit is negative. If equal amounts are placed in the three individual categories, it will lead to a loss of $A_{\text{LO}} - A_{\text{BL}} - A_{3C}$.

Case 3: The third case is that there are some contradictory bets that can be placed. A complex bet is contradictory if $P_{\text{Cbet}} < 0$ for any possible values of $x \in R$. This means that a bet like this would result a loss, regardless of the outcome of the spin, even if the spin comes in favor of the person who bets. For example, in an unlikely scenario of one placing straight-up bets with amount A for each of the 37 numbers, for all R numbers, then the profit will be

$$P_{\text{Cbet}} = 35A - 37A = -2A \quad (25)$$

It is to be noted that the above scenario exists only theoretically and just for demonstration purpose. The probability in winning complex bets is considered next. The probability of winning at least one simple bet is $P\left(\cup_{y \in Y} b_y\right)$. In calculating the probability of winning multiple simple bets when placed, some of these can be very simple numbers, while others can give rise to large mathematical and numerical expressions. The coverage of complex bet can be expressed as a union as $\frac{\#I}{\#R}$, where $\#I$ is the cardinality, that is the coverage of the complex bet I,

$$I = \bigcup_{y \in Y} b_y \quad (26)$$

For calculating the numerical values of these probabilities for specific complex bets, the number of elements of the union must be counted of the numerator. In the case that b_y are mutually exclusive, the cardinality or their union is as simple as

$$\#I = \frac{\sum_{y \in Y} \#b_y}{\#R} \quad (27)$$

If the sets b_y are not mutually exclusive, then the inclusion–exclusion principle can be used for their union as given under:

$$\left|\bigcup_{y=1}^{n} X_y\right| = \sum_{F\subseteq\{1,2,\ldots,n\}, F\neq 0} (-1)^{|F|-1}\left|\bigcap_{f\in F} X_f\right| |X_1 \cap X_2 \cap \ldots \cap X_n| \quad (28)$$

This expansion for inclusion–exclusion principle can help a player understand the coverage of their bet, but it does not conclusively indicate the probability of getting a non-negative profit.

Case 4: Complex bets of any kind, P_{Cbet}, are not constant on outcome x because of the very fact that the outcome could belong to more than 1 of the bets, b_i. This means that calculating the probability of getting a non-negative profit is not straightforward.

So, on a conquest to try generalizing the probability, the concept of a stair function is applied since P_{Cbet} follows a stair function.

A stair function is a kind of function, such that it is constant on each set belonging to a partition of its domain. In other words, it is a piecewise constant function. To show that P_{Cbet} is a stair function, a classic set theory theorem is made use of: If we consider S as an algebra of sets, $(b_y)_{y\in Y}$ is a finite family of sets from the algebra S, then there exists another finite family, $(c_w)_{w\in W,}$ of mutually exclusive sets from S such that

$$\bigcup_{y\in Y} b_y = \bigcup_{w\in W} c_w \quad (29)$$

For every $z \in Z$ we have, c_w is included in either one or more of the sets b_y. If $c_w \not\subset b_y$, then it is sure that $c_w \cap b_y = \phi$. To put it simply, for any finite family of sets, there is a partition of their union such that each set of that partition is included in at least one set b_y and is mutually exclusive with any other set b_y. A partition is a group of mutually exclusive sets that exhausts the initial set. A partition $\mathcal{M}$ is constructed such that it obeys the condition of the set theorem explained above. The family of sets $\mathcal{M}$ is built using the following steps.

There are two situations here:

1. If sets b_y have the property (the sets do not intersect any other of the rest)

$$b_y \cap \bigcup_{z\neq y} b_z = \phi \quad (30)$$

2. If sets b_y have the property (the sets that do intersect any one of more of the rest)

$$b_y \cap \bigcup_{z\neq y} b_z \neq \phi \quad (31)$$

It is observed that mutually exclusive cases are not taken up because of the simple fact that if sets $(b_y)_{y\in Y}$ are mutually exclusive, then $\mathcal{M} = (b_y)_{y\in Y}$.

To create a set of partitions, all non-empty intersections in the sets are considered first. Assuming that Π_n is a set with all n size intersections, where n would be the maximum number of intersections that meet each other. Starting off with the set Π_n, and all these sets can be directly added into the partition set $\mathcal{M}$. This is followed by

Π_{n-1}. If $\Pi_{n-1} \neq \phi$, all the sets from Π_{n-1} that do not have anything in common with the sets of Π_n must be taken (that is *n-1* sized intersections that do not include any *n* sized intersections), and each one of these sets in $\mathcal{M}$ is introduced, provided that they exist. Now, all the sets that do have something in common with Π_n are taken up. In case this is the situation, the sets such that parts that do not intersect have to be taken. So, let $H \in \Pi_{n-1}$ and $H_1, H_2, \ldots H_k \in \Pi_n$ such that $H_y \supset H, 1 \leq y \leq k$. Then, each set from the below is taken and added into the partition family $\mathcal{M}$.

$$H - \bigcup_{y=1}^{k} H_y \tag{32}$$

The above procedure is iteratively repeated until Π_2 is reached. After building a family of sets, the initially placed complex set is

$$\sum_{y=1}^{p} \mathcal{M}_y = I \tag{33}$$

where p refers to the number of elements in the partition family.

4 Specific Examples

Firstly, to enhance the understanding of complex bets, the probability of making a non-negative profit is taken up. With reference to Eq. 16, the same can be used if complex bets are disjoint. However, for a complex bet that is not disjoint, the profit of the bet will change very slightly. Instead of using the overall bet, the partitions as explained earlier are made use of, for realizing the profit function.

$$P_{\text{partition}} = \begin{cases} \sum\limits_{t \in A_y} W_{Ct} A_t - \sum\limits_{y \in R-(A_y, A_{y+1}, \ldots, A_{y+n})} A_y, & \text{if } x \in M \\ -\sum\limits_{y \in Y} A_y, & \text{if } x \in (R - M) \end{cases} \tag{34}$$

where $(A_y, A_{y+1}, \ldots, A_{y+n})$ are all the bets included in the partition. In short, this is done by adding the profit of the partition and subtracting whatever is not covered by the partition. Since the expectation of an event can be given in the form of a probability equation of the form,

$$p(\mathcal{F}_y) = \frac{\#\mathcal{F}_y}{\#R} \tag{35}$$

This is the probability of getting a number from the partition $\mathcal{F}_y$,

where $y \in Y$ and Y is a set of consecutive indexes, starting form 1 till the number of partitions in the partition set $\mathcal{M}$. The expected value of the profit of a complex bet can be found as below:

$$E(P_{\text{Cbet}}) = \sum_{\mathcal{F}_y \in \mathcal{M}} p(\mathcal{F}_y) * P_{\mathcal{F}(\text{partition})} = \sum_{\mathcal{F}_y \in \mathcal{M}} \frac{\#\mathcal{F}_y}{\#R} * P_{\mathcal{F}(\text{partition})} \tag{36}$$

The above equation can be simply used to find the expected value of the profit of a bet place, which can be positive or negative. The probability of getting a positive profit is the union of all expected profits when the profit of the partition is positive. Let those partitions of positive profits be $\mathcal{F}_1, \mathcal{F}_2, \mathcal{F}_3, \ldots, \mathcal{F}_r$. Since these partitions are obviously mutually exclusive, instead of taking a union, summation would suffice.

$$\sum_{u=1}^{r} \frac{\#\mathcal{F}_u}{\#R} \tag{37}$$

To illustrate the calculation of probability of positive profit, the following example is taken up:

$$I = \{(1D, 5), (HI, 7), (\text{split}\{18, 21\}, 3)\} \tag{38}$$

where 1D indicates first dozen, while HI denotes high numbers.

$$I = \left(b_y, A_y\right)_{y \in Y} \tag{39}$$

where A_y is the amount. The partitions are:

$$\mathcal{M} = \begin{cases} \mathcal{F}_1 = \{1, 2, 3, 4, 5, 6, 7, 8, 9, 10, 11, 12\} \\ \mathcal{F}_2 = \{19, 20, 22, 23, 24, 25, 26, 27, 28, 29, 30, 31, 32, 33, 34, 35, 36\} \\ \mathcal{F}_3 = \{21\} \\ \mathcal{F}_4 = \{18\} \end{cases} \tag{40}$$

Venn diagrams can be used as visual representation for making partition, and Fig. 3 depicts the Venn diagram for the above example. All the sets are mutually exclusive and exhaust the coverage of the bet I.

$$\begin{aligned} &\text{for } \mathcal{F}_1,\ P_{\text{partition}} = W_{C1}A1 - A2 - A3 = 10 - 7 - 3 = 0 \\ &\text{for } \mathcal{F}_2,\ P_{\text{partition}} = W_{C2}A2 - A1 - A3 = 7 - 5 - 3 = -1 \\ &\text{for } \mathcal{F}_3,\ P_{\text{partition}} = W_{C2}A2 + WC3A3 - A1 = 7 + 51 + -5 = 53 \\ &\text{for } \mathcal{F}_4,\ P_{\text{partition}} = W_{C3}A3 - A2 - A1 = 51 - 5 - 7 = 39 \\ &\text{in case } x \in R - (\mathcal{F}_1 \cup \mathcal{F}_2 \cup \mathcal{F}_3 \cup \mathcal{F}_4),\ P_{bet} - A1 - A2 - A3 = -15 \end{aligned} \tag{41}$$

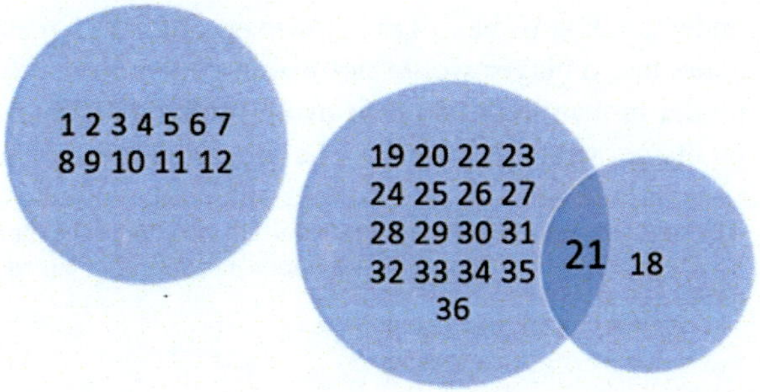

Fig. 3 Venn diagram for the example

As seen above, probability of a positive profit is shown in three cases, out of which two cases exhibit a very good probability.

$$
\begin{aligned}
p(\text{non negative profit}) &= \tfrac{\#\mathcal{F}_1+\#\mathcal{F}_3+\#\mathcal{F}_4}{\#R} = 37.84\% \\
p(\text{positive profit}) &= \tfrac{\#\mathcal{F}_3+\#\mathcal{F}_4}{\#R} = 5.41\% \\
p(\text{high profit}) &= \tfrac{\#\mathcal{F}_3+\#\mathcal{F}_4}{\#R} = 5.41\%
\end{aligned} \tag{42}
$$

If we just change the amount on the split bet to 2, probability values for $\mathcal{F}_1$ through $\mathcal{F}_4$ change to $1, 0, 36$, and 22, respectively. This scenario also increases the probability of non-negative and positive profit to 83.78% and 37.84%, respectively, while that of high profit stays the same at 5.41%, thus increasing the probability of overall profit.

5 Conclusion

This paper gave an overall account of how the mathematical modeling of probability and profit of single-zero roulette can help in understanding of bets. Firstly, the bets were put into forms of probabilistic events for building the model. Then grouping these into sets, many results from set theory were made use of to generalize bets. The results were then observed from the perspective of understanding the profit, given the outcome of a spin. The modeling for simple bet just gave a generalization and helped in understanding better. But this did not increase the probability of winning or not losing the game. This is easily shown from the results that the probability for simple bets did not depend on the amount bet. Also, the expectation of profit did not vary from bet to bet. Since the most prominent thing in gambling is that the amount bet is the only thing in the player's hands, this did not increase the probability of him/her winning.

Modeling complex bets, on the other hand, turned out to be extremely useful to enhance understanding of bets. The examples showed that the probability increased with the change in bet amount which meant that the modeling enhanced players' understanding of bets. This, however, did not suggest a winning strategy, nor guarantee that a player would not lose a game, but it just means that there is a way to model probability and profit using probabilistic events so that a player, depending on his/her strategy, can beforehand look at the bets he/she wants to place.

Acknowledgements The authors would like to thank the management of Birla Institute of Technology and Science Pilani, Dubai Campus, for offering their support and facilities in the completion of this work.

References

1. Donati, M.A., Chiesi, F., Primi, C.: A model to explain at-risk/problem gambling among male and female adolescents: Gender similarities and differences. J. Adolesc. **36**(1), 129–137 (2013)
2. Epstein, R.: The theory of gambling and statistical logic. Academic Press, New York (1977)
3. Malmuth, M.: Gambling theory and other topics. 6th edn. Two Plus Two, Florida (2004)
4. Croucher, J.S.: Gambling and sport: a statistical approach. Macquarie Lighthouse Press, Sydney (2003)
5. Barboianu, C.: Roulette odds and profits—the mathematics of complex bets. ISBN-10: 973-87520-7-8
6. Croucher, J.S.: A comparison of strategies for playing even money bets in roulette. Teach. Stat. **26**(3), 19–23 (2004)
7. Croson, R., Sundali, J.: The Gamblers's fallacy and the hot hand: empirical data from Casinos. J. Risk Uncertainty **30**(3), 195–209 (2005)
8. Small, M., Tse, C.K.: Predicting the outcome of roulette. Chaos 22(3) (2012)
9. Lichtenstein, S., Slovic, P., Zink, D.: Effect of instruction in expected value on optimality of gambling decisions. J. Exp. Psychol. **79** (2, Pt.1), 236–240 (1969)
10. Abuhanna, D.: Roulette: More than just a chance. Theses.27. University of Nevada, Las Vegas (2015)
11. Pflaumer, P.: A statistical analysis of the roulette martingale system: examples, formulas and simulations with R. In: 17th International Conference on Gambling and Risk Taking. May 27–30, 2019, Las Vegas, NV (2019).
12. Seal, K.C., Przasnyski, Z.H.: Illustrating probability through roulette: a spreadsheet simulation model. Spreadsheets Educ. **2**(1), 73–94 (2005)
13. Kavokin, A.V., Sheremat, A.S., Petrov, M.Y.: Bunching of numbers in a non-ideal roulette: the key to winning strategies. J. Math. Stat. Anal. **1**(1), 98–104 (2018)

Nonlinear Computational Crack Analysis of Flexural Deficit Carbon and Glass FRP Wrapped Beams

Tara Sen

Abstract Large numbers of reinforced concrete (RC) beams all over the world, especially in the earthquake prone regions in India, have undergone severe deterioration and damages to the extent that immediate resolutions in the form of structural retrofitting and rehabilitation is required for an extension in their respective service life and also for their continuity in functioning. The RC beams and their characteristic performances such as flexural strength can be suitably enhanced by the bonding of fibre reinforced polymer composites of carbon and glass. This paper deals with the nonlinear finite element computational analysis of the RC beams retrofitted using carbon and glass FRP, both before FRP bonding and after FRP bonding, for flexural characteristic studies, alongside a comprehensive comparison with the test data. The results obtained from the non-linear finite element analyses were corroborated with laboratory data. The comparison of the nonlinear finite element analysis with that of the experimental data revealed that computational analysis of structural components through nonlinear finite element analysis and its subsequent crack pattern study is an effective method of virtual analysis of structures.

Keywords Nonlinear analysis · Finite element analysis · Computational analysis · FRP · Flexure · Beams

1 Introduction

Enormous numbers of reinforced concrete (RC) beams built in our country have displayed lack of adherence to earthquake resistant design codes (IS-1893-2002), as a result of which various past earthquakes as well as other environmental conditions have rendered these structural components damaged or suitable case for rehabilitation. These structurally deficient components have further undergone deterioration due to various environmental affecting parameters working on them such as corrosion, affect of freeze and thaw, chloride and salt ingress, moisture absorption, sulphide

T. Sen (✉)
Department of Civil Engineering, National Institute of Technology, Agartala, Barjala, Jirania 799046, India

O. Castillo et al. (eds.), *Applied Mathematics and Computational Intelligence*, Springer Proceedings in Mathematics & Statistics 413,
https://doi.org/10.1007/978-981-19-8194-4_14

attack, etc. This alarming number of structurally deficient beams highlights the need for structural upgradation or retrofitting using cost effective measures and materials. Various experimental investigations have proved that the utilization of unidirectional or bi-directional woven fibre reinforced polymer composite (FRP) textiles made of carbon (CFRP), basalt (BFRP), glass (GFRP), natural (NFRP), and aramid (AFRP) fibres in their textile or woven form bonded with epoxies at specific locations or regions on the RC beams, have resulted in sufficient enhancement in load carrying capacities with proper environmental protections. These FRP bonding techniques have been successfully provided as a resolution for the up-gradation or retrofitting of various structural components. While there are large numbers of laboratory illustrations depicting the effectiveness of using FRPs for structural retrofitting or rehabilitation purpose, there is a lacuna in appropriate modelling, followed by nonlinear finite element-based computational analysis of these FRP wrapped RC beams, so that test data can be easily verified with virtual analysis. Enormous data in the virtual computational analysis can result in providing a strong database for quick and reliable analysis of these FRP bonded or wrapped structural components. Appropriate modelling, followed with quantified measurements of several parameters such as the load carrying capacities, deflections, stresses and strains, and the failure loads and the cracks in these FRP wrapped beams can provide a strong database in building the concept of computational analysis and crack study of FRP wrapped beams [1, 3–5, 9–11]. This paper deals with the nonlinear finite element analyses of the RC beams retrofitted using carbon and glass FRP, both before FRP bonding and after FRP bonding, and the flexural characteristic studies were done alongside a comprehensive comparison with the test data.

2 RC Beam Models and Its Description and Retrofitting Plans

Design of all the RC beams was carried out in accordance to IS-456: 2000. Beams were designed in such as manner that the beams undergo flexural failure first before the onset of shear failure. Table 1 describes the designations of all the beams that were modelled in the finite element tool, i.e. ANSYS 12.0, alongside their description and failure type. In each set, there were two beams, as shown in the table below:

Table 1 Design RC beams and retrofitting schemes

Beam designations	Description of the model	Failure type
Model CBF	Controlled RC beam model where no FRP will be bonded	Flexural Failure
Model BFC	RC beam model where carbon FRP will be bonded	Flexural Failure
Model BFG	RC beam model where glass FRP will be bonded	Flexural Failure

2.1 Finite Element Models

Six RC beams were modelled using the ANSYS 12.0 finite element method (FEM) software, as per the plan and the beam designations as described in Table1. In real time, i.e. in the laboratory, the retrofitting was carried out by hand lay-up systems with the wet lay-up technique, where the woven fibres were bonded to the beam surface for the formation of a FRP composite material. All the guidelines as prescribed by the manufacturers were followed for the retrofitting methods. Even in the software, during the modelling of BFC and BFG models, the FRPs of carbon and glass, respectively, were bonded to the respective RC beams in the U wrapping configurations, along the entire beam length in full wrapping technique using the inter-nodal connectivity feature in the computational analysis, as shown in Fig.1. The inter-nodal connectivity of all the FRP, loading plate, and RC beam model is shown in Fig.2. U wrapping technique of retrofitting or FRP bonding aids in flexural and shear load enhancement characteristics alongside increment in the overall structural confinement and ductility. Hence, three sided U wrapping in along the entire length of the beams was selected as the retrofitting or the strength enhancement plan.

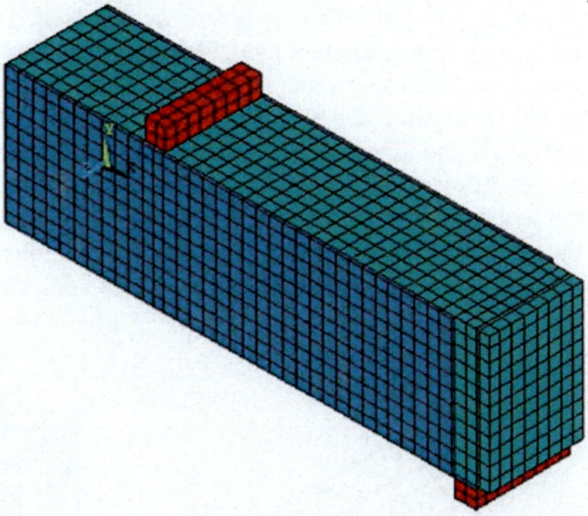

Fig. 1 FRP bonded on three sides of the RC beam FEM model, in U wrapping scheme

Fig. 2 Inter-nodal connectivity establishment between FRP, loading plates, and RC beam in the FEM model

3 Nonlinear Finite Element Computational Analysis of FRP Wrapped Beams

3.1 Brief Steps in Finite Element Computational Modelling

This part of the article deals with the modelling, evaluation, and successive result presentation of the nonlinear finite element computational analysis using ANSYS 12.0 tool. Only half of all the beams were modelled due to geometrically symmetric conditions. Since, discreet modelling was done, hence all components of the beams were modelled using separate discreet volumes, such as the concrete beams, steel plates for loading, and also the steel supports of the beam, all were modelled separately. The nonlinear finite element analysis of the controlled RC beams as well as the different retrofitted RC beams belonging to different retrofitting schemes, i.e. as designated and described in Table.1, were all numerically investigated by the nonlinear finite element method computational method of analysis, with the following brief steps, and the entire pictorial description is given in Fig. 3.

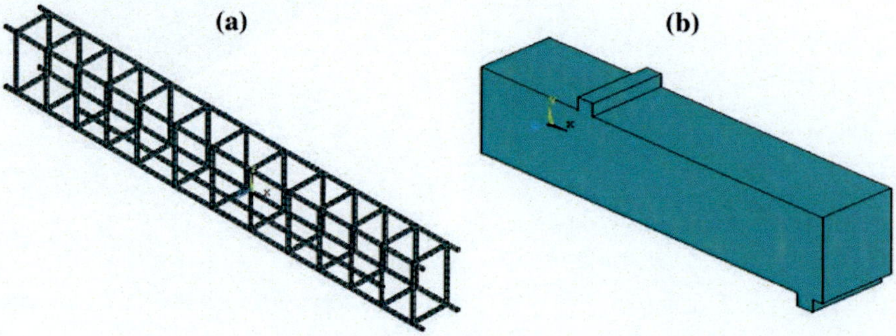

Fig. 3 FEM model for reinforcements **a** for flexural failure **b** Boundary of the entire FEM model with "edge-only" option

1. A general formulation of nonlinear finite element model using the ANSYS 12.0 tool, incorporating the modelling of the reinforced concrete beams, with discreet modelling of the concrete and the reinforcement, the FRP composites, the steel supports and the loads, and the entire model being subjected to two-point loading system was presented.
2. Suitable finite elements were considered, and SOLID65 for concrete, LINK8 for reinforced steel, SOLID45 for steel plates for loadings and supports, and finally, SOLID layered46 for the FRP composites were utilized for building up the entire structure. Mapped meshing was carried out with the suitable elements for building up the entire structural model. Elemental nodal connectivity was established through the mapped meshing. The concrete solid 65 element used here is based on the Von Misses failure criterion which is again dependent on the Willam and Warnke model, which has been basically used to define the failure surface of the concrete [15].
3. Suitable displacement boundary conditions were assigned to the entire structural model, so as to constrain the model for obtaining the unique solution. The symmetric boundary conditions on the axis of symmetry and the support boundary conditions for modelling of the roller support were considered. The force applied at each node on the plate utilized for assigning load to the RC beam was P divided by the total number of nodes selected, for the actual force applied, the force P was applied in the direction of UY.
4. Small displacement and static nonlinear finite element analysis was carried out utilizing the solution and controls options. Also suitable load step options were incorporated for the finite element model.
5. The finite element modelling was capable of predicting failure for concrete materials, i.e. both cracking and crushing failure modes were accounted for. Finally, after the nonlinear finite element analysis, all the crack patterns were obtained in ANSYS 12.0, at the corresponding ultimate stages of loadings.
6. The study of all the crack patterns obtained using the nonlinear finite element analysis by ANSYS 12.0, revealed good agreement with the experimentally obtained results in terms of the crack patterns and the fracture modes displayed by all the specimens.

3.2 Finite Element Model Representations and Build-up In ANSYS 12.0

Solid65 element was used for meshing along with all the properties of all the concrete [7, 16]. The mapped meshing option was used in the ANSYS 12.0 tool, since mapped meshing uses a large number of straight edged sides for meshing, hence the meshing is uniform throughout the model and is best suited for generation of inter-nodal connectivity between all elements in the finite element model (FEM) [8]. Discreet modelling was carried out with the inputs being provided separately for all the different elements in the model such as concrete, reinforcements, FRP composite sheets, and steel plates for loading and support, after the discreet modelling was complete, the overlapping or gluing of all the discreet volumes was done. The "edge option", in ANSYS 12.0 tool, with the sub-option "element outlines for non-contour/contour plots followed by edge-only", when used, must deliver the entire outline boundary of the whole model. If the result of edge option is just the outline boundary of the entire model, then this assures us that the entire model will act as one unit and also assures us the concept of node sharing between adjacent sub-elements. Figure 3a is the FEM model for reinforcements used to aid in flexural failure, and Fig. 3b is the outline or boundary edges of the FEM model with the "edge-only" option, which displays the entire FEM model as one unit. Figure 4 presents the symmetrical boundary conditions used on the axis of symmetry of the RC finite element model beam. Figure 5 represents the point and line of load application on the FEM beam. Figure 6 presents the displacement restraints in order to build the support constrains in the FEM beam.

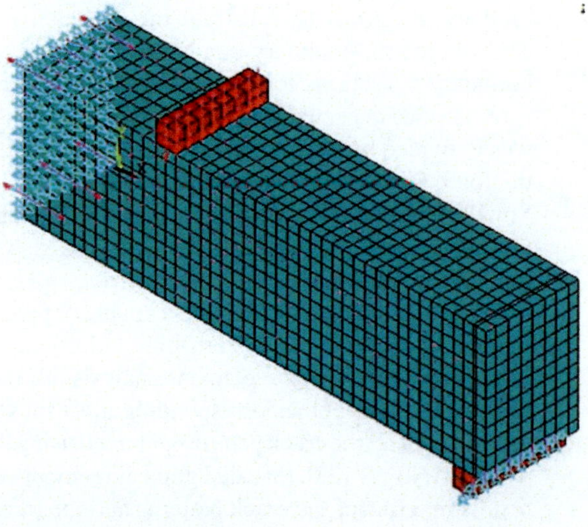

Fig. 4 FEM model displaying symmetrical boundary conditions on the axis of symmetry

Fig. 5 Load application through the steel loading plates in the FEM model

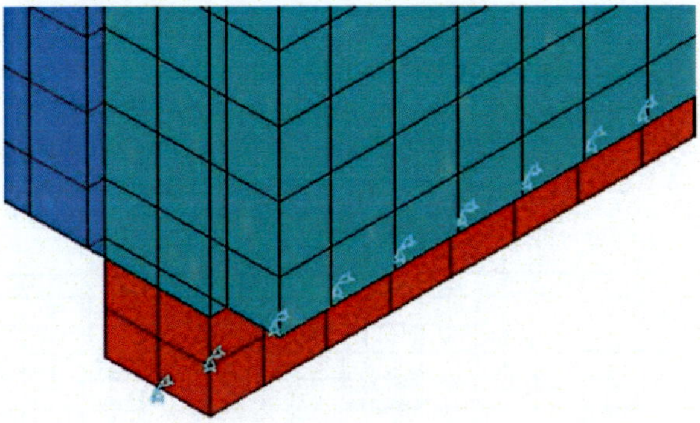

Fig. 6 Displacement constraints in the steel support

4 Flexural Crack-Crush Computational Studies

Large literature is available which deals with the cracking and crushing behavioural failure pattern study of concrete in nonlinear computational analysis [2, 6, 12–14]. The red lines basically indicate concrete cracking having small crack width, whereas the blue-green basically represent excessive concrete cracking with larger crack widths in the concrete zone. In the CBF beam (controlled RC beam), vertical longitudinal cracks appeared in the beam middle zone, i.e. the pure flexure zone of the beam. Along with the vertical flexure cracks, some flexural-shear cracks at the end of the pure bending zone were also present. Both longitudinal cracks and slightly 45 degree inclined flexure-shear cracks were seen in the CBF model. The computational nonlinear static analysis displayed an ultimate flexural load carrying capacity of 86 KN. The BFC model in the computational analysis displayed extensive flexural cracks, i.e. by flexural cracks which were tensile cracks in nature. Vertical flexural cracks alongside some flexural-shear cracks were observed in the flexure zone of the BFC model. The location of blue-green marks which represent excessive concrete cracking can be seen in the centre of the model, mainly in the middle zone, this was due to sudden catastrophic rupture of the carbon FRP, followed by FRP de-bonding from the mid-span face of the BFC model beam. The computational nonlinear static analysis displayed an ultimate flexural load carrying capacity of 226 KN. The BFG model in the computational analysis too displayed extensive flexural cracks, i.e. by flexural cracks which were tensile cracks in nature. Vertical flexural cracks alongside some flexural-shear cracks (diagonal tensile cracks in nature) were observed in the flexure zone of the BFG model. The location of blue-green marks which represent excessive concrete cracking can be seen in the location where the FRP area ends in the model near the supports, mainly in the beam end zones, this was due to sudden catastrophic de-bonding of the glass FRP from the face of the BFG model beam. The computational nonlinear static analysis displayed an ultimate flexural load carrying capacity of 196 KN.

Both the cracking and crushing features were considered in the modelling of the beams. All the figures presented basically display the contour plot of the displacement vector sum and the crack patterns exclusively at the failure load (flexural ultimate load) of the corresponding samples. Figure 7 presents the diagonal tensile flexural cracking in the FEM models in (a) controlled beam (CBF), (b) RC beam wrapped with carbon FRP (BFC), and (c) RC beam wrapped with glass FRP (BFG).

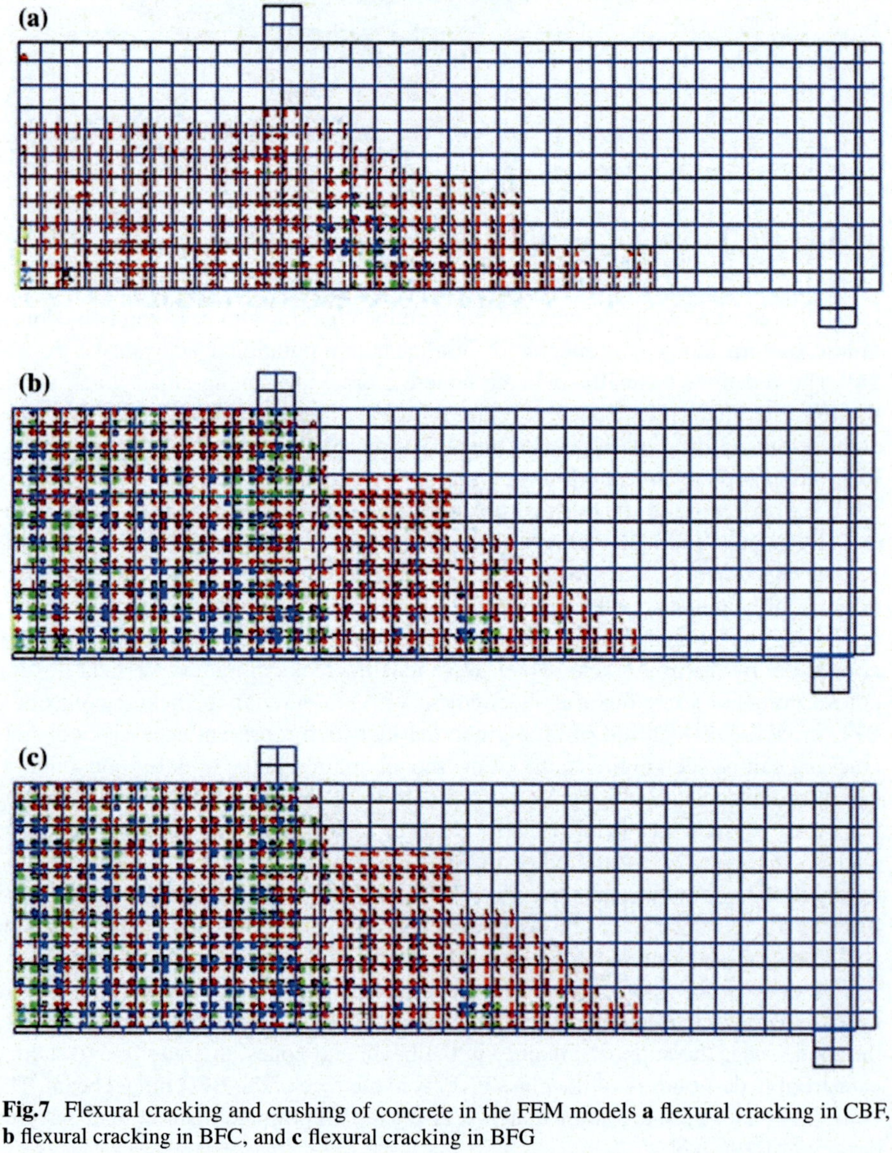

Fig.7 Flexural cracking and crushing of concrete in the FEM models **a** flexural cracking in CBF, **b** flexural cracking in BFC, and **c** flexural cracking in BFG

Table 2 Design RC beams and retrofitting schemes

Beam Designations in FEM analysis	Ultimate flexural load carrying capacity in computational analysis	Ultimate flexural load carrying capacity as obtained through laboratory test data	PLaboratory/PComputational	Percentage difference
Model CBF	86	80	0.93	8
Model BFC	226	200	0.88	13
Model BFG	196	180	0.92	9

Mean0.912
Standard deviation0.029
Coefficient of variation0.032

5 Conclusions

In terms of quantifying the data variation spread, the standard deviation is the most utilized measure. The standard deviation is the typical or average distance a value is to the mean of all the data. A value of higher standard deviation implies that the data is more spread out and therefore has more variability. The coefficient of variation is a measure of dispersion of a probability distribution. It is defined as the ratio of the standard deviation to the mean. Table 2 gives a comparison of the ultimate flexural load carrying capacities as achieved from the laboratory experimental data analysis and that obtained from the nonlinear finite element computational static analysis, for all the beam models. The flexural load carrying capacity was found to be in close agreement with that obtained from experimental test data followed by analysis.

References

1. Al-Rousan, R., Haddad, R.: NLFEA sulfate-damage reinforced concrete beams strengthened with FRP composites. Compos. Struct. **96**, 433–445 (2013)
2. Demakos, C.B., Repapis C.C., and Dapacities enhanced rivas, D.: Investigation of Struc- tural Response of Reinforced Concrete Beams Strengthened with Anchored FRPs. Open Constr. Build. Technol. J. **7**,146–157 (2013)
3. Hawileh, R.A.: Nonlinear finite element modelling of RC beams strengthened with NSM FRP rods. Constr. Build. Mater. **27**, 461–471 (2012)
4. Hawileh, R.A., El-Maaddawy, T.A., Naser, M.Z.: Nonlinear finite element modelling of concrete deep beams with openings strengthened with externally-bonded composites. Mater. Des. **42**, 378–387 (2012)
5. Hawileh, R.A., Naser, M.Z., Abdalla, J.A.: Finite element simulation of reinforced concrete beams externally strengthened with short-length CFRP plates. Composites: Part B **45**, 1722–1730 (2013)

6. Jindal, A.: Finite element modelling of reinforced concrete exterior beam-column joint retrofitted with externally bonded fiber reinforced polymer (FRP). Master's Thesis, Department of civil engineering, Thapar University, Patiala , Punjab (2012)
7. Kachlakev, D., and Miller, T.: Finite element modelling of reinforced concrete structures strengthened with FRP laminates Final Report SPR 316, Oregon department of transportation. For Oregon Department of Transportation Research Group200 Hawthorne SE, Suite B-240Salem, OR 97301–5192 and Federal Highway Administration, 400 Seventh Street SW, Washington, DC 20590 (2001)
8. Luo, Y., Li, A., Kang, Z.: Parametric study of bonded steel–concrete composite beams by using finite element analysis. Eng. Struct. **34**, 40–51 (2012)
9. Özcan, D.M., Bayraktar, A., Sahin, A., Haktanir, T., Türker, T.: Experimental and finite element analysis on the steel fiber-reinforced concrete (SFRC) beams ultimate behavior. Constr. Build. Mater. **23**, 1064–1077 (2009)
10. Omran, H.Y., El-Hacha, R.: Nonlinear 3D finite element modelling of RC beams strengthened with prestressed NSM-CFRP strips. Constr. Build. Mater. **3**, 74–85 (2012)
11. Padmarajaiah, S.K., Ramaswamy, A.: A finite element assessment of flexural strength of prestressed concrete beams with fiber reinforcement. Cement Concr. Compos. **24**, 229–241 (2002)
12. Potisuk, T., Higgins, C.C., Miller, T.H., and Yim, S.C.: finite element analysis of rein- forced concrete beams with corrosion subjected to shear. Adv. Civil Eng. Volume 2011, Article ID 706803, p. 14 (2011)
13. Santhakumar, R., Chandrasekaran, E.: Analysis of Retrofitted reinforced concrete shear beams using Carbon fiber composites. Electron. J. Struct. Eng. **4**, 66–74 (2004)
14. Saifullah, I., Hossain, M.A., Uddin, S.M.K., Khan, M.R.A., Amin, M.A.: Nonlinear analysis of RC beam for different shear reinforcement patterns by finite element analysis. Int. J. Civil Environ. Eng. **11**(01), 63–74 (2011)
15. Willam, K.J., Warnke, E.P.: Constitutive model for triaxial behaviour of concrete. In: Seminar on Concrete Structures Subjected To Triaxial Stresses. In: International Association of Bridge and Structural Engineering Conference, Bergamo, Italy, pp. 174 (1974)
16. Wolanski, A.J.: Flexural behavior of reinforced and prestressed concrete beams using finite element analysis. Master of Science thesis, Faculty of the Graduate School, Marquette University, Milwaukee, Wisconsin (2004)

Economic Benefit Analysis by Integration of Different Comparative Methods for FACTS Devices

Rituparna Mitra, Sadhan Gope, Arup Kumar Goswami, and Prashant Kumar Tiwari

Abstract In modern power system, dependence on power electronic devices are mostly affected by voltage sag and results process interruption and that leads to huge monetary loss. Voltage sag is a bottleneck of streamlining power quality. FACTS devices are implemented with an eye to restoration of quality power supply. DVR and STATCOM are well known devices for the avoidance of voltage sag. Optimum selection of device for voltage sag mitigation in terms of economy is the objective of this proponent. Net present value (NPV), payback method, internal rate of return (IRR) and profitability index (PI) have been surfaced to critically analyze the acceptable device with least cost with higher return. Mathematical equations of these comparative indexes have been dealt with to give a concrete idea and growing an opinion to select the coveted device for practical purpose. The rampage of voltage sag can thus be tamed and quality of distribution can be achieved.

Keywords FACTS device · Net present value (NPV) method · Payback method · Internal rate of return (IRR) method · Profitability index (PI) · Cost–benefit analysis

1 Introduction

Different power quality problems, i.e., voltage sag, voltage swell, transients, harmonics, etc., are the determined by voltage quantity or magnitude variation [1]. A distribution system consumer generally faces around 70 events per year of voltage

R. Mitra (✉)
Swami Vivekananda University, Kolkata, West Bengal 700121, India
e-mail: rituparnamitra1990@gmail.com

S. Gope
Mizoram University, Aizawl, Mizoram 796004, India

A. K. Goswami
National Institute of Technology, Silchar, Assam 788010, India

P. K. Tiwari
Motilal Nehru National Institute of Technology, Allahabad, UP 211004, India

O. Castillo et al. (eds.), *Applied Mathematics and Computational Intelligence*,
Springer Proceedings in Mathematics & Statistics 413,
https://doi.org/10.1007/978-981-19-8194-4_15

sag 0.7 p.u. [2, 3]. Any industrial process consists of variable speed drives, and they are prone to voltage sag. Failure of protective schemes in distribution system and voltage sag condition incurs enormous economic loss in the industry. To overcome this situation, economic analysis of voltage sag mitigation devices is very rational. The conventional method like NPV, payback, IRR and PI of management studies have been involved in this article to put to devices under scanner for selection of the favorable one. Cost–benefit study is a far reaching analysis to give appropriate result by consideration of initial cost, equipment lifetime, scrap value, maintenance cost, running cost, depreciation of the devices and ultimately to patronize the benefit and allocation in a particular distribution system. Industrial customer, utilities and end users are the best judge to accept most economical power quality improvement device to maintain hassle free situation. This concept is a priori intimation to the user to choose the effective device. Placement of the device also an integrated part of cost–benefit situation. Power transmission and distribution, utility and customers helps aptly to handle and resolve optimal decision. Comparative study is the cognizable decision making factor in this article. The benefit derived is minimum loss due to power quality issues in the system. End user pokes into the matter of power quality cost to the tune of loss in production, revenue cost, scrap cost, wages and overtime charges for labor, and replacement or recovery charges [4]. Types of load for different customers are the decisive factor for selection of voltage sag mitigation device. A few number of mitigation device have been selected economically and technically for power quality improvement. Sustenance of power quality improvement, device installation and maintenance cost have to be looked into. Prior to installation of any device, system operator must look into cost–benefit analysis for continuation of improvement in power quality. Individual approach of different end user to be monitored with an angle to advanced technology and improved design of control scheme. The backdrop study of the application is required to be done before implementation. Power quality cost borne by the consumer and their desire to pay for are matter for consideration. Let there be comparison between power quality cost and power quality improvement cost. Actually power quality cost consists of loss in production, equipment trip, wastage of materials, etc. Power quality improvement cost includes installation and maintenance cost due to improvement of power quality device. Electricity supplier has the privilege to select multiple customers to bear power quality improvement cost in economic analysis. With adopted voltage sag mitigation scheme if more than one consumer gets benefit, then the economic analysis has been considered as feasible and smart and named as "premium power park". Choosing different alternatives for voltage sag mitigation is an economical task. This task has been usually done by facility manager and utility engineer. Costs of power quality and performance improvement for different alternatives are the ultimate goal of the task. The most feasible alternative depends on the cost of the problem and total operating cost of different available solutions [5]. Adequate knowledge on power quality is essential to arrange power quality problem and their solutions chronologically according to their importance [6]. Recently, in market, different types of voltage sag mitigation devices are available. Some of them are used generally and regularly where some of them are installed in special cases. These devices effectively reduce number of process interruption due to voltage sag, and costs of these devices are variant with types. Voltage sag mitigation cost varies with the scheme used for mitigation in the system. Therefore, it is more difficult to choose effective voltage sag mitigation device with minimum cost in technical and economical point of view. The most optimal option is low cost sag mitigation device with long service period (lifetime investment). Investment cost of voltage sag mitigation devices prevents outage cost and subject to cost–benefit analysis. Cost of the voltage sag mitigation devices is a prime factor for selection of them. Different devices and technologies must be applied to analyze and compare their effectiveness for alleviation of voltage sag after installation. Reduction of total power quality cost justifies the economic point of voltage sag mitigation [7] in terms of cost of the devices and reduction in downtime cost bear by the consumer. An opportunity cost is usually essential between two cost mechanisms. Power quality costs depend on the severity of power quality issues. Installation of voltage sag mitigation devices has two parts: (a) Fixed Cost: It includes cost of mitigation device, installation, labor cost and allocation cost. (b) Variable Cost: It includes operating cost of the voltage sag mitigation devices. It also includes losses due to overheating, maintenance and replacement costs of the parts of device, cooling cost, etc. Maintenance cost and other losses are included in annual percentage of purchase cost. The comparison among all the alternatives for performance improvement is determined by annual cost of individual alternatives. That annual cost comprises of power quality variation cost and annual solution cost. The costs considered for optimal device selection are device installation cost, outage cost not eliminated by sag mitigation equipment and maintenance cost. Some available standard financial models are generally used to assess different alternatives, e.g., payback period, net present value (NPV) and internal rate of return (IRR) [8]. Huge capital investment projects are typically dealt with NPV method. Payback method is most usual method to assess any process. Payback method calculates the time required to recover initial investment in a business. Only if payback period result crosses a threshold value defined by authority, the business is eligible to prosecute. Typically, the threshold value for payback period for any project is 2–4 years. This method is substandard than NPV method as it considers any type of cash flow after payback time. Positive cash flow during first five years of establishment is bestowed in payback method. With this method, number of process interruption before and after immunity device installation determination is also possible. The annual profit can be premeditated by multiplying diminished number of process interruption by the cost of one process interruption. The net benefit of a process is calculated by comparing different annual profits of different voltage sag mitigation strategies. The objective of this article is to minimize annual cost. Techno-economical solution as a means of minimum total power quality cost point is the final solution.

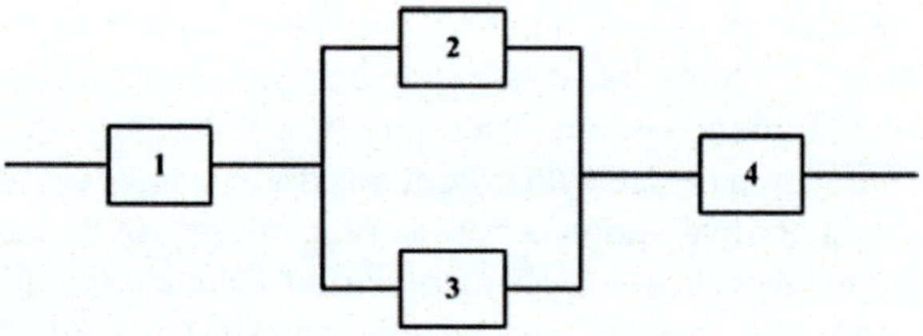

Fig. 1 General configuration of sensitive equipment in a process

2 Number of Process Interruption Due to Voltage Sag

Voltage sag occurs due to symmetrical and unsymmetrical faults. Residual phase voltages are determined with definite fault rates of distribution lines determined by interval type 2 fuzzy system [9] and fault type and locations are selected by monte-carlo simulation [10–13]. The voltage sag magnitude depends on the position of fault occurrence. Single line to ground fault, double line to ground fault, triple line fault and double line fault have been considered for calculation of voltage sag per year. The lowest magnitude of all the phases are considered for voltage sag determination in case of unsymmetrical faults [12]. Generally, it is observed that number of shallow voltage sag occurrence is more than deep voltage sag. An industrial process must consist of sensitive equipment like PLC, ASD and computers in different series–parallel combinations [17]. Protective devices of the process determine voltage sag clearing time. It is assumed that all faults can be cleared by primary protection scheme [14–16]. An example of process with different equipment (1: PLC, 2: ASD, 3: PC and 4: AC contactor) is shown in Fig. 1.

In general way, the chances of trip to the process can be written as

$$C_{Trip} = 1 - \left[\prod_{j=1}^{m} \left(1 - \prod_{k=1}^{n} c_{j,k} \right) \right] \tag{1}$$

where m: Total number of series connected equipment, n: Total number of parallel connected equipment and $c_{j,k}$: Cumulative probability of jth equipment in kth series.

3 Discussion on Economic Strategies

Traditionally, corporate sectors use well reputable methods for economic scrutiny of a project named as net present value (NPV) method, internal rate of return (IRR) method, payback method and profitable index, etc.

3.1 Net Present Value (NPV) Method

Possibility of investment profitability can be calculated accurately with the help of net present value method. The basic norm of finance is "a dollar today is worth more than a dollar tomorrow, because the dollar today can be invested to start earning interest immediately" [18]. The present value for an overdue payoff is equal with actual payoff multiplied with discount factor. The value of discount factor is considered less than 1. If discount factor is more than 1, then today's value of dollar will be more than tomorrow's value. If is expected payoff for one year, then

$$\text{Present Value } (PV) = \text{Discount Factor} \times R_1 \tag{2}$$

$$\text{Discount Factor} = 1/1 + r \tag{3}$$

r: Rate of return

Discount factor reciprocates today's investment in terms of future return. Rate of return is a reward demanded by investors for receiving deferred payment. Accurate present value can be calculated since expected payoff is reduced by rate of return obtainable by equivalent investment substitutes in the capital market. Rate of return can be called as discount rate or hurdle cost or opportunity cost of capital. This return can be called as opportunity cost as it means return disavowed from other alternative when another alternative is chosen. NPV method considers all the cash flows during the lifetime of a project. Cash flow in the following years is calculated with discounted rate named as opportunity cost of capital (OCC) or hurdle rate or discount rate or required rate of return. This cost is calculated from initial time ($t = 0$) and results into present value of the cash flows. The least rate of return putative for an assured project is known as opportunity cost of capital. Net present value of a project is the difference between discounted cost cash flow and discounted benefits. Only positive valued NPV of a project is acceptable as the initial investment fully returns with benefits [19]. For a probable investment in a project, there is no certain methods available in economics to determine opportunity cost of capital. The method of traditional finance and the method of modern finance are the most usual approaches to calculate OCC [20]. In modern method of financial analysis, risk management is considered while decision making. Here, traditional method has been used to enumerate different voltage sag mitigation strategies. It analyzes and helps to make decision considering investment in voltage sag mitigation method. Traditionally, the base to calculate OCC is the rate named weighted average cost of capital (WACC) to placate basis of capital. Projects with rate of return more than WACC is acceptable as it will be able to return money more than capital expenses of that project. Sources of capital with equity and debt are considered to calculate WACC as follows [20]:

$$\text{WACC} = (\% \text{ in equity} \times \text{cost of equity}) + (\% \text{ in debt} \times \text{cost of debt}) \tag{4}$$

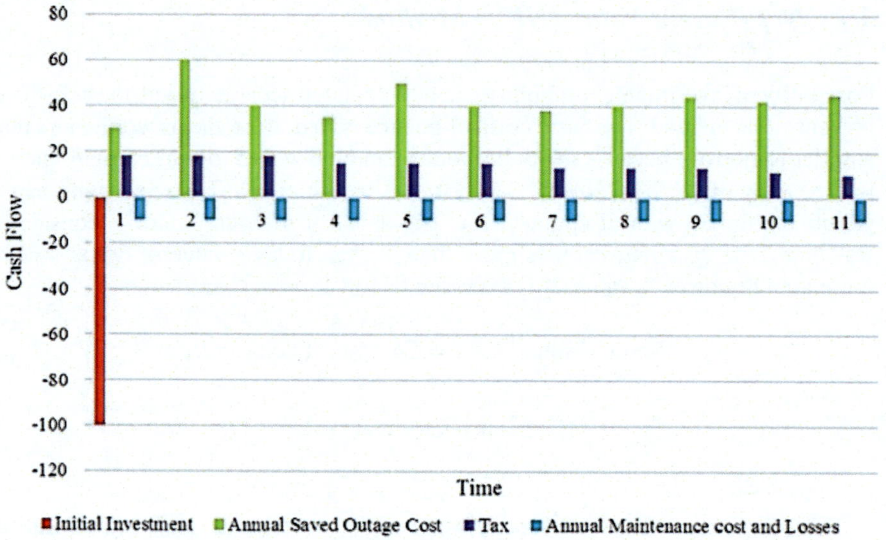

Fig. 2 Cash flow in voltage sag mitigation approach

Cost of debt is calculated as after-tax basis and this is multiplication of cost of debt and (1-tax rate) [18]. Investment in voltage sag mitigation method to arrest process interruption due to voltage sag is subject to cost–benefit analysis. Cost–benefit analysis has been done using NPV method [18]. Positive value of NPV of the discounted cash flow rationalizes investment in mitigation equipment. Figure 2 represents cash flow in voltage sag mitigation equipment investment.

The cash flow includes initial investment, maintenance and standby cost, tax and outage costs. Neglecting salvage value of the equipment, NPV of cash flow has been calculated as

$$NPV = \sum_{t=0}^{n} \frac{CF_t}{(1+r)^t} - C_0 \tag{5}$$

where CF_t: Net cash flow at time t, C_0: Initial investment, r: Discount rate, t: Time duration in years and n: life time of the investment. Every business has predefined cost of capital. With cost of capital and project life time, if NPV returns a positive value, then only the project is acceptable.

3.2 Payback Method

Several industries do not use NPV method for their investment [21]. Large capital investment companies generally use NPV method, and those companies must follow one of the following characteristics:

- Investment value more than 200,000 euro
- High-ranking decision making
- Presence of finance company or investor's influence
- Business and financial committee takes the decision and does financial evaluation.

Generally, payback or discounted payback methods are used as decision tool for maximum companies. The required time to get initial investment back (payback time) is the outcome of this approach. The approach is only feasible if payback time is lower than the defined threshold value of the company. Typically, the defined threshold value is 2–4 years. In discounted payback method, interest rate of discounted cash flow for $t = 0$ is been considered. Payback method is esteemed for its easy application, but it is not efficient as NPV method. Payback time is the time required for a project to breakeven point. Generally, payback time is calculated in months. Payback time can be calculated as the following equation:

$$\text{Payback(months)} = {}^{\text{NetInvestment}}/_{\text{NetAnnualReturn}} \times 12 \tag{6}$$

Here, net investment includes mitigation equipment and installation cost. Net annual return is the difference between annual expenses and annual benefits. Many industrial customers demand for less or equal to 1–2 year as payback time for their invested project. This is equal to 50–100% return [21, 22].

3.3 Internal Rate of Return (IRR)

IRR method contemplates all marginal cash flow thru investment of the project like NPV method. IRR is a discount rate at which NPV value is zero. If a project has more internal rate of return from its predefined hurdle rate, then only the project is acceptable. IRR is the discount rate for which present value of the cash flow in a project is equal with its initial investment. Mathematically, IRR can be expressed as

$$\sum_{t=0}^{n} \frac{CF_t}{(1+R)^t} - C_0 = 0 \tag{7}$$

R: Internal rate of return.

Projects with higher IRR than cost of capital are only acceptable [22, 23].

3.4 Profitability Index (PI)

This is a ratio term and its ratio between project benefit in the present time to present cost value. It is expressed as

$$PI = PV/ - C_0 \tag{8}$$

where PV: Discounted future cash flow (forecasted) and C_0: Initial investment at $t = 0$.

A project is acceptable only when PI is more than 1. This criterion is same as NPV method as PI of 1 is same as NPV of 0. With NPV method, profitability indexing also gives same result for a project acceptance and rejection. But the drawback of this method is same with IRR method that for mutually exclusive event this method is unsure to take decision [22, 23].

4 Illustrations

Wise decision making for voltage sag mitigation requires sufficient data for power quality finance analysis. Adaptation of logical and systematic process for economic scrutiny is important for selection of utmost cost effective resolution for power quality problem.

The steps are analyzed with the following case studies to validate the approach.

Step 1:

System Power Quality Analysis: In this first step, occurrence of different types of disturbances in different system buses and their frequency of disturbance are considered. Voltage sag is the most influencing power quality issue on operation of system. Voltage sag assessment is the primary part of economic analysis, descriptively discussed in [9]. Occurrence of symmetrical and unsymmetrical faults in different buses and lines are analyzed to calculate residual voltage of the system [24, 25]. The analytical method has been applied to determine voltage sag performance in bus 12 of Barak valley distribution system as shown in Fig. 3.
Step 2:

Estimate Power Quality Variation Cost: Installation of mitigation device requires enormous money. So, financial losses due to process interruption and number of process trip estimation are required to be calculated a priori. The financial loss due to voltage sag without installation of any voltage sag mitigation device is shown in Table 1. High sensitive equipment for topology I and moderate sensitive equipment in topology II faces maximum financial losses. To scrap voltage sag most efficiently, the service provider or utility tries to reconfigure the system with available re-closer or switches with available information regarding voltage sag

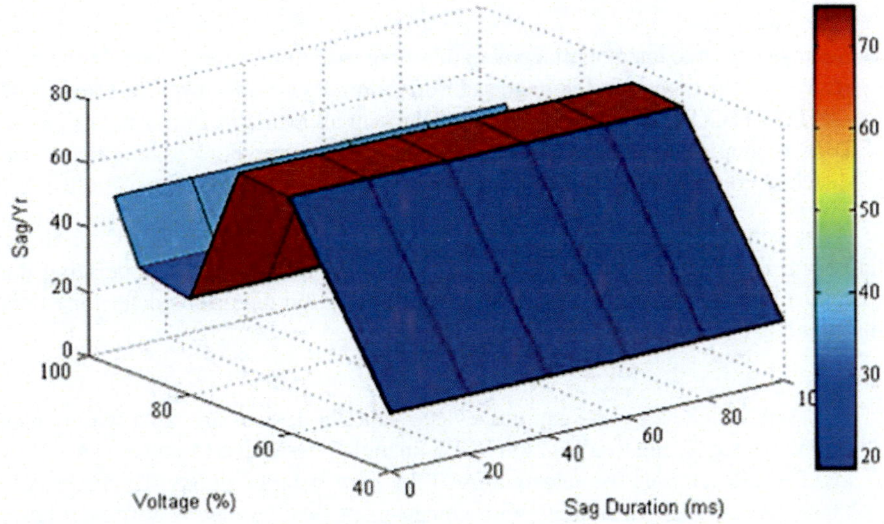

Fig. 3 Voltage sag performance at bus 12

performance and financial losses faced due to this. After voltage sag mitigation device installation, the savings in different cases are discussed here.

Case study I:

Savings with the placement of FACTS device: Single voltage sag mitigation device has been installed in bus number 12 in Barak valley distribution system. For higher and moderate sensitive equipment, topology I and II have been chosen, respectively. Reduction of financial loss after placement of mitigation device is shown in Table 2.

Table 1 Financial losses (in Cr.) due to voltage sag without mitigation device

Topology	I	II	III	IV	V	VI
High sensitive	0.50	0.35	0.33	0.24	0.48	0.34
Moderate sensitive	0.08	0.25	0.11	0.11	0.03	0.13

Table 2 Improvement in financial losses (in Million.) after mitigation device installation

Sensitivity	Without mitigation device	With DVR		With DSTATCOM	
		Financial loss	Benefit	Financial loss	Benefit
High (topology I)	50.109	45.5	4.609	47.6	2.509
Moderate (topology II)	25.0545	22.8	2.2545	23.02	2.0345

Case study II:
Savings with two unequal placement of voltage sag mitigation device: In this case study, two different sized voltage sag mitigation device have been selected. DVR and DSTATCOM are the chosen device among all available options. The size of DVR is smaller than DSTATCOM. These two mitigation devices are placed with different combinations, and the resulted cost saving is mentioned in Table 3.

After voltage sag mitigation device, the amount of reduction in financial loss has been determined. Now, it is turn to calculate installation cost, operating and maintenance cost of the devices. The cost of mitigation devices is determined and discussed below.

Step 3:
ost of Voltage Sag Mitigation Devices: FACTS devices are most widely used as voltage sag mitigation tool to reduce financial losses due to voltage sag. DVR and DSTATCOM are the selected FACTS device installed in bus 12. The price of these two devices is calculate with equations of FACTS device cost calculation and depicted in Table 4C.

From [26], equations to determine costs $(C_{STATCOM}, C_{DVR})$ of FACTS device are mentioned. The only variable in that equation is the ratings $(S_{STATCOM}, S_{DVR})$ of those devices.

$$C_{STATCOM} = 2.80 \times 10^3 S_{STATCOM}^2 + 217.9 \times S_{STATCOM} + 24.28 \tag{9}$$

$$C_{\mathrm{DVR}} = 2.25 \times 10^3 \times S_{\mathrm{DVR}}^2 + 189.5 \times S_{\mathrm{DVR}} + 47.21 \tag{10}$$

Assumption: discount rate (r) 12% for these devices, the rating of these devices is 25 MVAR, 1 \$ = 70.35 rupees on 19.05.2019. Cost of STATCOM is 123.49 in million rupees, and cost of DVR is 99.26 million rupees. Generally operating and maintenance cost are some percentage of initial investment. In this case, running cost of DVR is 10% and STATCOM is 15% of its initial investment. Therefore, net investment for DVR is 109.186 million rupees and for STATCOM is 142.0135 million rupees. For 5 MVAR capacities, the net investment on DVR and STATCOM are 4.43 and 5.75 million rupees.

Step 4:

Economic Analysis: After calculating voltage sag mitigation cost, device cost along with operating cost, way-out of economic analysis are implemented in the following segment.

A. Net Present Value:
Corporate sector finds NPV method as most efficient tool to decide regarding investment. In this case, NPV is calculated for 20 years. Both high and moderate

Table 3 Financial losses (in Million.) after mitigation of two mitigation devices

Sensitivity	Without mitigation device	Type I		Type II		Type III		Type IV	
		Financial loss	Benefit	Financial loss	Benefit	Financial loss	Benefit	Financial loss	Benefit
High (topology I)	50.11	48.44	1.67	46.86	3.24	48.15	1.96	47.08	3.03
Moderate (topology II)	25.05	22.64	2.41	21.04	4.02	22.23	2.83	21.34	3.71

Table 4 Price range of DVR and STATCOM [26]

Operating range (MVAr)	$/KVAr			
	DVR		STATCOM	
	Min	Max	Min	Max
100	60	100	90	130
200	42	78	70	110
300	40	70	64	98
400	40	60	60	100

sensitive equipment in process are considered with installation of either FACTS device here from customer's perspective. For both moderate and highly sensitive process, the mitigation device (either DVR or STATCOM) has been placed in bus 12. Initially, all present value has been calculated and NPV analysis is shown in Fig. 4 and 5 for placement of mitigation device. These two figures clearly depict irrespective of the equipment sensitivity, positive cash flow starts only after second year of the installation. NPV method is more acceptable for moderately sensitive equipment than highly sensitive one. At the end of 20 years, total return is 10.17 after STATCOM placement and 2.577 after DVR placement.

Comparing the above two figures, it is concluded that initial investment in DVR is lesser than STATCOM, still STATCOM is more attractive option for voltage sag mitigation irrespective of equipment sensitivity as it returns more money in the

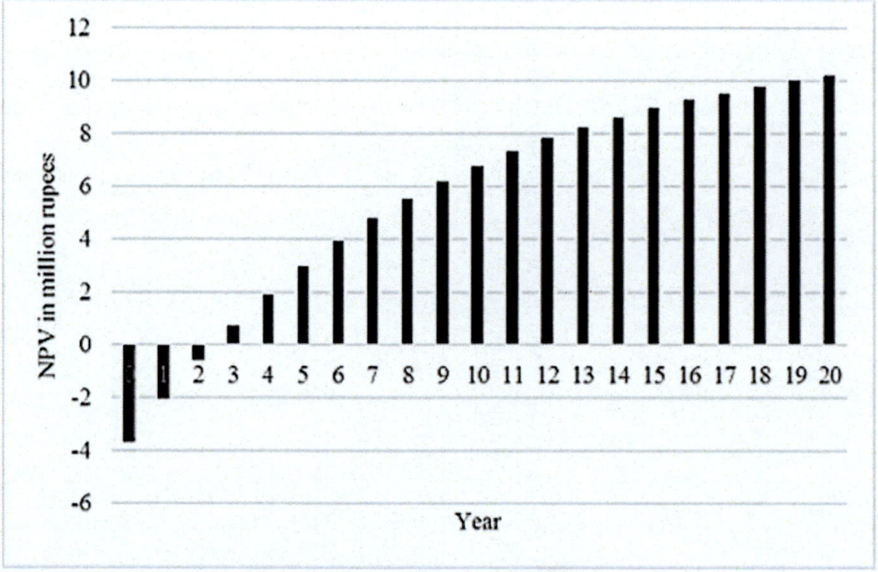

Fig. 4 NPV values of STATCOM for paper industry

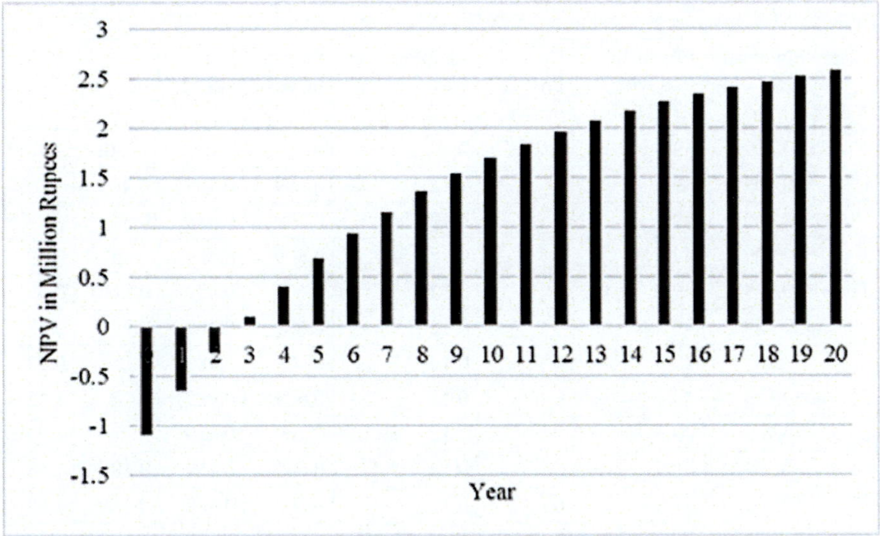

Fig. 5 NPV values of DVR for paper industry

long run. This decision is solely based on utility perspective. And from customer's perspective, STATCOM is more useful for high sensitive equipment and DVR is optimal for moderately sensitive equipment. Detailed NPV analysis for DVR and STATCOM and cash flow after every year for high and moderate sensitive equipment for paper industry in Barak valley distribution system is shown in Tables 5 and 6.

Table 5 NPV analysis of installation of STATCOM (in million rupees)

Yr	Financial loss	Financial loss with STATCOM	Cost of FACTS device	Maintenance cost	Saving in loss	Net saving	PV	NPV
0	50.109	47.6	3.69	0	0	0	−3.69	−3.69
1	50.109	47.6	0	0.66	2.509	1.849	1.65	−2.04
2	50.109	47.6	0	0.66	2.509	1.849	1.47	−0.57
3	50.109	47.6	0	0.66	2.509	1.849	1.32	0.75
4	50.109	47.6	0	0.66	2.509	1.849	1.18	1.93
5	50.109	47.6	0	0.66	2.509	1.849	1.05	2.98
6	50.109	47.6	0	0.66	2.509	1.849	0.94	3.92
7	50.109	47.6	0	0.66	2.509	1.849	0.84	4.76
8	50.109	47.6	0	0.66	2.509	1.849	0.75	5.51
9	50.109	47.6	0	0.66	2.509	1.849	0.67	6.18
10	50.109	47.6	0	0.66	2.509	1.849	0.6	6.78

Table 6 NPV analysis of installation of DVR (in million rupees)

Year	Financial loss	Financial loss with STATCOM	Cost of FACTS device	Maintenance cost	Saving in loss	Net saving	PV	NPV
0	50.109	49.5	1.09	0	0	0	−1.09	−1.09
1	50.109	49.5	0	0.12	0.609	0.489	0.44	−0.65
2	50.109	49.5	0	0.12	0.609	0.489	0.39	−0.26
3	50.109	49.5	0	0.12	0.609	0.489	0.35	0.09
4	50.109	49.5	0	0.12	0.609	0.489	0.31	0.4
5	50.109	49.5	0	0.12	0.609	0.489	0.28	0.68
6	50.109	49.5	0	0.12	0.609	0.489	0.25	0.93
7	50.109	49.5	0	0.12	0.609	0.489	0.22	1.15
8	50.109	49.5	0	0.12	0.609	0.489	0.2	1.35
9	50.109	49.5	0	0.12	0.609	0.489	0.18	1.53
10	50.109	49.5	0	0.12	0.609	0.489	0.16	1.69

Payback Method: In corporate sector, payback method is been used as decision tool. Using Eq. (6) the payback period can be calculated. The payback period for STATCOM is 8.2 months and for DVR is 6.7 months. As customer's perspective DVR has faster recovery rate than STATCOM. Payback period is directly connected with the sensitivity of the equipment. For high sensitive equipment, payback period is lesser for DVR, but for moderate sensitive equipment, STATCOM has faster recovery rate. While two mitigation device is considered together, then the payback period is18 months, 10.5 months, 21.6 months and 9.81 months for type I, II, III and IV, respectively. Type IV installation type recovers the investment quickly among the four.

Internal Rate of Return: Either of the mitigation device has been placed for assessment of IRR. IRR is positive for all the investment. That reflects the investment in feasible and acceptable. The IRR for STATCOM is little more than DVR irrespective of equipment sensitivity and placement of the device. Placement of STATCOM is more beneficial in this case, but the method is sensitive with equipment sensitivity.

Profitability Index: Single voltage sag mitigation device has been considered for this approach. It is evident from the result that STATCOM has profitability index more than 1 and DVR's profitability index is less than 1. The profitability index varies with equipment sensitivity.

The overall result of economic analysis is shown in Table 7.

From this result, it can be concluded that in NPV method positive cash flows after 2 years of installation of voltage sag mitigation device and STATCOM is more beneficial for power quality improvement in this case. As IRR for STATCOM is higher than DVR, it is again opted for most feasible solution. Profitability index for both the equipment is almost same. As corporate sector mainly believes in NPV method,

Table 7 Economic analysis of installation of FACTS devices

Devices	Payback period (month)	NPV (million rupees)	IRR (%)	PI
STATCOM	8.2	10.17	38	3
DVR	6.7	2.577	35	3.1

the result obtained from NPV method is acceptable for this case. So, STATCOM is the most suitable device for voltage sag mitigation, and it is efficient in long run.

5 Conclusion

NPV method as a key factor gives higher score to STATCOM in relation with higher return of investment cost than DVR. Since initial investment of DVR is less than STATCOM, the payback period supports DVR for less payback period. Under IRR method, the rate of return of STATCOM is higher than that of DVR because initial investment for DVR is lower but cash flow for STATCOM is higher. Profitability index of both the devices is same. Under this review, STATCOM defeats DVR by 2:1 index results. This result is based on mathematics and reputed indexes to give a concrete and clear picture for identification of the FACTS device in any system conducive for effective management of power distribution system.

References

1. Steciuk, P.B., Redmon, J.R.: Voltage sag analysis peaks customer service. IEEE Comput. Appl. Power **9**, 48–51 (1996)
2. Divan, D., Sutherland, P., Grant, T.: Dynamic sag corrector: a new concept in power conditioning. Power Qual. Assurance 42–48
3. Hietpas, S.M., Naden, M.: Automatic voltage regulator using an AC voltage-voltage converter. IEEE Trans. Ind. Appl. **36**(1), 33–38 (2000)
4. Barry, W.: Kennedy. McGraw-Hill, Power Quality Primer (2000)
5. McGranaghan, M., Roettger, B.: Economic evaluation of power quality. IEEE Power Eng. Rev. **22**(2):8–12 (2002)
6. Didden, M.: Techno-economic analysis of methods to reduce damage due to voltage dips. Ph.D. dissertation, Katholieke Universiteit Leuven, December 2003.
7. Pereira, F.C., Souto, O.C.N., De Oliveira, J.C., Vilaca, A.L.A., Ribeiro, P.F.: An analysis of costs related to the loss of power quality. In: Proceedings 8th International Conference on Harmonics and Quality of Power, vol. 2, pp. 777–782 (1998)
8. Von Jouanne, A., Banerjee, B.: Assessment of voltage unbalance. IEEE Trans. Power Delivery **16**(4), 782–790 (2001)
9. Mitra, R., Goswami, A.K., Tiwari, P.K.: Voltage sag assessment using type-2 fuzzy system considering uncertainties in distribution system. IET Gener. Transm. Distrib. **11**(6), 1409–1419 (2017)
10. Information about cement manufacturing (2013) [online]. Available : http://www.essroc.com.
11. Qader, M.R., Bollen, M.H.J., Allan, R.N.: Stochastic prediction of voltage sag in a large transmission system. IEEE Trans. Industr. Appl. **35**(1), 152–162 (1999)

12. Park, C.H., Jang, G.: Stochastic estimation of voltage sags in a large meshed network. IEEE Trans. Power Delivery **22**(3), 1655–1664 (2007)
13. Juarez, E.E., Hernandez, A.:An analytical approach for stochastic assessment of balanced and unbalanced voltage sags in large system. IEEE Trans. Power Delivery, **21**(3):1493–1500 (2006)
14. Olguin, G., Bollen, M.H.J.: The method of fault positions for stochastic prediction of voltage sags: a case study. In: International Conference on Probabilistic Methods Applied to Power Systems. Naples, Italy, pp. 22–26 (2002)
15. de C. Filho, J.M., de Abreu, J.P.G., Ju´lia, H.A., Noronha, C.C.: Analysis of power system performance under voltage sags. Electr. Power Syst. Res. **55**:211–18 (2000)
16. Milanovic, J.V., Gupta, C.P.: Probabilistic assessment of financial losses due to interruptions and voltage sags-Part II: Practical Implementation. IEEE Trans. Power Delivery **21**(2), 925–932 (2006)
17. Milanovic, J.V., Gupta, C.P.: Probabilistic assessment of financial losses due to interruptions and voltage sags-part i: the methodology. IEEE Trans. Power Delivery **21**(2), 918–924 (2006)
18. Brealey, R., Myers, S.: Principle of corporate finance, 6th edn. McGraw- Hill, New York (2003)
19. Didden, M.: Voltage disturbances-considerations for choosing the appropriate sag mitigation devices. In: Power Quality Application Guide-Copper Development Association (2005)
20. Chambers, D., Lacey, N.: Modern corporate finance, theory and practice. Addison Wesley Educational Publisher inc (1999)
21. Baggini, A., Bua, F.: Investment analysis for PQ solutions. Leonardo power quality initiative (LPQI) UK., [online]. Avialable: http://www.lpqi.org./lpqi_archieve_contribute//2_5_low.pdf?6074/l094136091_658_658_CONTRIBUTE_PATH_2_5_Low.pdf
22. Thasananutariya, T., Chatratana, S., Mc Granaghan, M.: Economic evaluation of solution alternative for voltage sags and momentary interruptions. Electr. Power Qual. Utilization. Mag. **1**(2), 17–26 (2005)
23. Didden, M., Stockman, K., D'Haeseleer, W., Belmans, R.: Cost-benefit analysis of solutions protecting industrial process with variable speed drives against voltage sags. Electr. Power Qual. Utilization Mag. **1**(2), 11–16 (2005)
24. Juarez, E.E., Hernandez, A.: An analytical approach for stochastic assessment of balanced and un balanced voltage sags in large systems. IEEE Trans. Power Delivery **21**(3), 1493–1500 (2006)
25. Mitra, R., Goswami, A.K., Tiwari, P.K.: Optimal selection of voltage sag mitigating devices for micro-level customer in distribution system. IET Renew. Power Gener. **13**(1), 191–200 (2019)
26. IEEE Recommended Practice for Monitoring Electric Power Quality. IEEE Standard 1159. New York (1995)

Optimal Pricing with Servicing Effort in Two Remanufacturing Scenarios of a Closed-Loop Supply Chain

Ashish Kumar Mondal, Sarla Pareek, and Biswajit Sarkar

Abstract In today's competitive markets, only retailer is not focused on selling products in the retail market. The manufacturer focuses on product remanufacturing that can fulfill the customer's demand and increase manufacturer's revenue earnings. This study optimizes the selling price and remanufactures products with a similar commitment as to the fresh one. The urge for remanufactured products highly delineates their price and carbon emission reduction, as these two features are most important to customers for indicating the value and standard of the renovating products. This research explores a classical optimization process under a closed-loop supply chain management with declining the emission of CO_2 considering carbon tax and selling price-dependent market demand. Here, Stackelberg game theory is utilized to solve the model with distributor and manufacturer both. Both distributor and manufacturer remanufacture together, where they sell new and remodeled products. The analytical solution gives the optimum selling price with reduced carbon emission.

Keywords Supply chain management · Remanufacturing · Third-party logistics · Carbon tax · Technology sharing

1 Introduction

In today's supply chain, it leads to a challenging situation to maintain a balance between environmental issues and economic issues. The carbon tax restricts carbon

A. K. Mondal · S. Pareek
Department of Mathematics and Statistics, Banasthali Vidyapith, Banasthali, Rajasthan 304022, India

B. Sarkar (✉)
Department of Industrial Engineering, Yonsei University, 50 Yonsei-ro, Sinchon-dong, Seodaemun-gu, Seoul 03722, South Korea
e-mail: bsbiswajitsarkar@gmail.com

Center for Transdisciplinary Research (CFTR), Saveetha Dental College, Saveetha Institute of Medical and Technical Sciences, Saveetha University, 162, Poonamallee High Road, Velappanchavadi, Chennai, Tamil Nadu 600077, India

O. Castillo et al. (eds.), *Applied Mathematics and Computational Intelligence*, Springer Proceedings in Mathematics & Statistics 413,
https://doi.org/10.1007/978-981-19-8194-4_16

emission from the manufacturing section in a closed-loop supply chain management (CLSCM) to decrease carbon emission. The manufacturer involves in remanufacturing instigates a carbon tax scheme with dual tax payments on manufacturing and remanufacturing levels. This work is done in the following process: first, take an optimal strategy to store products. After that, consider carbon tax to decrease CO_2 and earn more profit in the supply chain by manufacturing new and used products. It helps to find out the values of decision variables. Further, a CLSCM with reduced CO_2, exchange strategy, quality improved product, carbon tax, and selling price-dependent market demand of products is shown under remanufacturing scenarios for supply chain performance. The manufacturer and distributor remanufacture returned products where the third-party logistics (3PL) is liable for gathering used products from the customers. In this case, the manufacturer gives a technology license to the distributor to remanufacture a portion of the collected used products. The remainder portion is taken back to the manufacturing company for remanufacturing. Here, Stackelberg game is used for decision making to optimize the supply chain's profit. Different researchers have constructed different models where price-dependent demand is analyzed or the customer's attraction for green products. This study helps to answer the research queries mentioned below:

- Which supply chain is the best among two remanufacturing structures?
- How does carbon emission related with the remanufacturing of the supply chain?
- How can the selling price affect the demand for products in the market?
- What is the relation between the return strategy, carbon emission, and quality improvement effort in supply chain management?
- How does the technology license sharing impact supply chain management (SCM)?

The remaining portion is equipped as follows: Sect. 2 is introduced for a brief description of a similar effort in literature. Secondly, Sect. 3 is presented for problem description, notation, and assumptions. Next, in Sect. 4, the mathematical model is introduced. After that, Sect. 5 consists solution methodology and Sect. 6 consists numerical example. Managerial insights are discussed in Sect. 7. At last, in Sect. 8, the conclusion of the research is penned.

2 Literature Analysis

Some authors focused on product quality development's and suggested coordination agreements without not considering different collecting and remanufacturing propositions for greening. This research concentrates on the selling price-dependent market demand, quality improvement effort, remanufacturing, and carbon tax for carbon emission declination to instigate customer's requirement.

2.1 Selling Price-Dependent Demand

Market demand is important in the supply chain to fulfill the business goal. Generate a sufficient amount of market demand through a traditional policy is getting tough day by day due to increase use of internet. Amidst the variables, the most important is the buyer's income, quality, the price of related products, the standard of the brand, and the choice of customers. Xu et al. [28] explored the deterministic model by introducing the Stackelberg game theory to meet market demand disruption. Teunter and Van der Laan [26] analyzed the remanufacturing process where the demand depends on the initial returning rate. Yadav et al. [19] analyzed a production process for deteriorating products but without discussing a remanufacturing process or a hybrid manufacturing process for a selling price-dependent demand. Chaudhari et al. [17] discussed a payment policy for a supply chain model with a selling price-dependent demand. A study was inferred to prove the inverse proportionality of selling price with market demand [2].

2.2 Carbon Tax

Lower emission of CO_2 is an essential subject in recent research. Ji et al. [11] inspected a coupling retailer's inventory strategy to decrease CO_2, tax, cap, and trade for greening. Moreover, they considered different methods with supporting examples. Zhao et al. [32] analyzed CO_2 policies to decrease the cost of advanced techniques in two-stage SCM. They set up low CO_2 commodities with more expenses. Qin et al. [15] established an inventory model with environmental concerns and credit interval. They explored the greening influence on the inventory model under the tax for CO_2 and cap and trade strategy. Kugele and Sarkar [18] examined different carbon emission reduction policies from a system with carbon tax. Datta [4] developed some different environmental inventory model to detect optimum greening profit for decreasing emission by carbon tax strategy without scarcity. A carbon tax is an alternative approach to deal with climate change [6]. Allan et al. [1] examined that greenhouse gas emission is decreased for an adequate carbon emission tax in northern Europe, Taiwan, and Scotland. Still, an reduction tax has a negative result on the economy. Liu et al. [13] presented an alternative path for utilizing the income from CO_2 tax. They proved that subsidizing carbon tax and introducing new technique extend the spread for low carbon economy. In this current model, carbon tax for emission reduction is introduced in a CLSCM to optimizing profit between two scenarios.

2.3 Supply Chain Management

Giovanni [7] scrutinized the relation between a fixed refund and variable refund impacts. Customers' decisiveness to build an exchange argued that the maximum pays off in CLSCM by fixed abatement scheme is an ideal Markov solution under random insistence conditions. When a manufacturing company producing new products, then random exchange rates and manufacturing defects describe a severe role in exchanging remaining products. Giri and Sharma [8] traditionally estimated the exchange policies, by taking an exchange strategy to give back used products and compensate them. Xu et al. [29] analyzed the give back policy variously to demonstrate that the manufacturing company and retailers can collaboratively complete the exchange strategy, regardless of the individual or way. In this study, depending upon the analysis of CLSCM, product design, manufacturing, distributing, and reconstructing are developed. An application under various backgrounds and constraint was accomplished for optimizing the total cost of CLSCM by Demirel et al. [5]. They studied the problems of multi-product CLSCM and utilized mixed-integer linear programming to resolve the end-of-life vehicle recycling process. Yang et al. [30] studied an optimizing network stability grown in a CLSCM structure, which depends on different discrimination theory. Savaskan et al. [21] wrote a survey on three steps for the renewing convey of CLSCM, in which manufacturer's recycling, distributor's remanufacturing, and 3PL's remanufacturing were explained. Savaskan and Wassenhove [20] imposed a game theoretical structure to find benefit of the manufacturing company to remanufacture alone with recycling over competing distributors. Jacobs and Subramanian [10] considered the procedure to deliver remanufacturing responsibility amidst the supply chain participants to boost the payoff. Chen and Chang [3] proposed a cooperative and other strategies to examine the conditions by which an original equipment manufacturing company can take an interdependent proposal for remanufacturing. Shi et al. [22] discussed the optimization model for the market value of new and waste products in CLSCM. Taleizadeh et al. [23] developed the consequences for participants' maximum benefits in different CLSCM structures and planned tariff arrangement in two directions. Ramani and Giovanni [16] established participants' activity in CLSC and considered other competitive games by the profit-generating commitment to collaborating firms. Assuming the distributor's concern, Zeng et al. [31] analyzed the optimum decisions and benefits under pentagonal non-cooperative and cooperative game structures, which proved the process for the maximum payoff in a coordination agreement. This specified coordination had shown the possible deal of the CLSCM by stimulating backgrounds and the competitive quality of the participants in CLSCM.

2.4 Research Gap

Several kinds of research have considered many structures where the requirements depend on the products' market value or reduction of carbon emission, and carbon tax on the supply chain performance. This study examines a selling price-dependent demand, carbon tax for emission reduction, and servicing effort in a production system under a game theoretical environment. Table 1 presents the work of earlier authors to define the research gap. This paper is an extension of Taleizadeh et al. [24]. Different remanufacturing scenarios are derived in the present study. This study considers carbon tax for low carbon emission, quality improvement, and selling price-dependent demand to addresses the question mentioned in the introduction

Table 1 Summarizes augmentation and absorption of authors. Author's contribution according to the research gap

Author(s)	Carbon tax	Remanufacturing	Stackelberg game theory	Selling price-dependent demand	Return policy	Quality improvement	Green supply chain
Savaskan et al. [21]	–	✓	✓	✓	✓	–	–
Wang et al. [27]	–	–	✓	–	✓	–	–
Chen and Chang [3]	–	–	✓	–	–	–	✓
Taleizadeh et al. [23]	–	–	✓	✓	–	✓	–
Taleizadeh et al. [24]	–	–	✓	✓	✓	–	–
Harris and Ogbonna [9]	–	✓	✓	–	✓	✓	✓
Liu et al. [12]	✓	✓	✓	–	✓	–	✓
Taleizadeh et al. [25]	–	✓	✓	–	–	✓	✓
Liu et al. [13]	–	✓	✓	–	✓	✓	✓
This paper	✓	✓	✓	✓	✓	✓	✓

section. Taleizadeh et al. [25] analyzed the performance of a CLSCM where two carbon reduction options like investing money in the emission reduction procedure together with a trade and cap regulation are introduced. Here, this model distinguishes the supply chain's optimal decision procedure and inspects the results in two remanufacturing cases. Here, Stackelberg game is used to analyze the model.

3 Problem Description, Notation, and Assumptions

The problem description is defined to mention the way of research. Similarly, notation and assumptions are defined to determine the problem with the appropriate format.

3.1 Problem Description

In this research, a CLSCM is considered, where a manufacturing company, distributor and a third party are involved for remanufacturing products. This model is considered for motivating the customers to buy the remanufactured items. This paper considers two remanufacturing approaches like E and F to establish several decision variables. For case E, the manufacturer and distributor return products in the form of high quality. In this case, customers get payment (j) per unit return product from the distributor. The manufacturer then receives the fees for a license (l) from the distributor to remanufacturer x units of the returns. The remaining portion of return products $(1-x)$ is sent to the manufacturer by the distributor for remanufacturing and distributor receives b per unit returned product from the fabricator. For case F, the third-party is added to collect the used products in supply chain. Here, the manufacturer give technology licenses to a third-party for remanufacturing. The third-party remanufacturers x units of the returned products and the manufacturer remanufacturers the remainder portion $(1-x)$. In this study, a carbon tax is introduced along with price-dependent demand for the popularity of products. Here, Stackelberg game is used to find optimum selling price, tax for carbon, and to optimize partners' profits in the supply chain.

3.2 Notation

Parameters	
c_m	Production cost per new manufactured product ($/unit)
c_1	Manufacturer's remanufacturing cost per unit ($/unit)
c_2	Unit remanufacturing cost of the distributor ($/unit)
c_3	Unit remanufacturing cost of the 3PL ($/unit)
T_1	Money saving from unit remanufactured product obtain by the manufacturer ($/unit)
T_2	Money saving from the remanufactured product for each unit obtained by the distributor ($/unit)
T_3	Money saving of the 3PL from unit remanufactured product ($/unit)
x	Portion of exchange product which distributor or 3PL remanufacturers
l	Manufacturer received fees for technology license of each unit of item from the 3PL or the distributor ($/unit)
w	Wholesale price of items per unit from the manufacturer ($/unit)
μ	Fixed portion of the exchange items (unit)
c	Influence positive coefficient of the recompense for exchange items ($/unit)
y	Coefficient of the rate of quality for exchange items
η	Unit carbon emission rate (grams/unit)
G_v	Government permit to each firm for carbon emission (grams)
P_v	Price per unit carbon emission by cap and trade policy ($/grams)
M	Produced quantities per order (units)
$x(.)$	PDF (probability density function)
$X(.)$	CPDF (cumulative probability distribution function)
G	Stands for Stackelberg game for the market leadership of the manufacturer
F, Y, H, S	Denote the manufacturer, distributor, 3PL, and supply chain, respectively
$p_{\max}$	Maximum selling price per unit product ($/unit)
$p_{\min}$	Minimum selling price per unit product ($/unit)
Decision	**Variables**
v	Reduced carbon emission rate
j	Distributor or 3PL give unit recompense to consumer for exchange items ($/unit)
k	Quality improvement
I	Distributor's expected inventory function (units)
p	Average selling price of the product ($/unit)
s	Service variable
Performance	**Indicator**
π_S	Profit of the supply chain ($)
π_F	Profit of the manufacturer ($)
π_Y	Profit of the distributor ($)
π_H	Profit of the 3PL ($)

3.3 Assumptions

The assumptions described below are provided to understand the developed models.

1. A single type of product is involved for different economic and environmental constraints under the remanufacturing scenarios. The manufacturer does not support the stock, low costs, and does not accept returned amounts to avoid trifling cases, which is the goal of this research (Taleizadeh et al. [24], Wang et al. [27]).
2. There is a condition $c_m \leq w \leq p$ which must hold for both the manufacturer and the distributor such that their profit margin remains positive. To ensure success with the return strategy, the payback mandatory to meet. As the factories cannot reduce all emissions, the carbon emission reduction rate (v) follows $0 \leq v \leq 1$. The decreasing rate of CO_2 is zero when the carbon emission is not reducing by the factories [27].
3. Only a portion of the return products meets their manufacturing ability and the remanufacturing is done by the downstream members such that the 3PL and the distributor. The manufacturing ability (x) is determined by an manufacturing technique, the remanufacturing scale, and the exchange deterioration products. When $x = 0$, it indicates that the downstream members are not a recycling participants, and when $x = 1$, it implies downstream members are recycled all products [29].
4. To maximize every member's profit, a Stackelberg game is used to obtain competitive decisions. This paper leads that manufacturer plays the dominant role, and the downstream members pursue him. Reversely, the manufacturers optimize the result determiners by the participation of downstream members.
5. The parallel work was done by Chen et al. [3]; they considered that original and remanufactured products are the same. They can motivate the customers to choose remanufactured products equally as the new products at new prices [24].
6. For exploring a remanufacturing scenario, it should be confirmed that the manufacturing cost of new products is higher than the remanufacturing price of every exchange product that is $c_1 < c_m$, $c_2 < c_m$, and $c_3 < c_m$ [27]. Selling price-dependent market demand is introduced here.
7. Carbon tax is constructed by the manufacturer and distributor's effort levels in scenario *E*. The carbon tax function is $p_v[\eta(1 - v)M - G_v]$. The manufacturer constructs the carbon tax and 3PL's effort levels in scenario *F*. Each item constructs the function of a carbon tax which is $p_e[\eta(1 - v)M - G_v]$.

4 Mathematical Model

In this model, the stochastic demand function is considered. For variable demand of the products in the market, various decision variables like selling price of products, quality improvement, carbon emission reduction, the cost to buy return products, and servicing efforts are considered. In many research, the cost of product and demand are in an inversely proportional relationship, but the relation between product and demand price is proportional in this present study. The partners of the CLSCM are manufacturer, distributor, and 3PL, respectively. For maximizing the supply chain's profit, it is needed to increase the market demand for products. The demand function is elaborated as

$$D(e, r, q, p, s) = a_0 + a_1 \frac{p_{\max} - p}{p - p_{\min}} + a_2 v + a_3 v^2 + a_4 j + a_5 j^2 + a_6 k + a_7 k^2 + a s^{\gamma} + \epsilon \tag{1}$$

in which a_0 is the potential market size, which does not depend on selling price (p), the decreasing rate of CO_2 emission (v), the used product return cost (j), and improving quality (k). Furthermore, a_1 is the positive coefficient of the selling price, a_2 and a_3 are the coefficient for the positive effect of the low carbon emission rate, a_4 and a_5 are the coefficient for the positive impact of the return price, a_6 and a_7 are the coefficient for the positive effect of quality improvement effort, a is scaling parameter for service, γ is the shape parameter for service, which gives positive impact in market urge, and over the uniform distribution function in $[0, k_1]$, ϵ is a random market demand. Money savings by remanufacturing per unit product by the manufacturer (T_1), by the distributor (T_2), and by a 3PL (T_3) are shown below.

$$T_1 = C_m - C_1 \tag{2}$$

$$T_2 = C_m - C_2 \tag{3}$$

$$T_3 = C_m - C_3 \tag{4}$$

The investment cost for quality improvement of the products is introducing a technique for decreasing emission of CO_2 along with accumulating the exchange commodities and tax for CO_2 ejection and, increasing the price of the items. This strategy negatively affects the market demand for the product. For finding return quantity of used products, we establish a relation between the price of return and improve quality of goods, which is shown in Eq. (5).

$$R(k, j) = \mu + cj - yk \tag{5}$$

in which parameter μ is a constant portion of the return used product which is not dependent on compensated cost and improvement quality of the products, c is the coefficient which gives positive effects for refund price on the return products, and y is the coefficient which gives adverse effects for improved quality on the return products.

4.1 Manufacturer's Mathematical Model

In this CLSCM, the manufacturer is a market leader, but the products do not sell without a distributor. The manufacturer helps the distributor to renew the used products in better quality. The manufacturer does not support the stock, and low costs do not accept return amounts to avoid trivial cases. There is a condition $c_m \leq w \leq p$ which must hold for both the manufacturer such that their profit margin remains positive. The manufacturer's optimal profit is calculated by subtracting all his costs from the earned revenue.

Carbon emission reduction cost

The cost is investing to decrease the carbon emission along the manufacturing period, which is the most essential for low carbon emission. For motivating customers to purchase environment-friendly products, investment is needed for this purpose which is shown by

$$C(v) = \theta \tfrac{v^2}{2}, \tag{6}$$

where θ is the investment coefficient, which is not changing for decreasing the CO_2 emission rate.

Cost for quality improvement of the product

This cost function for improving the quality of products is shown by

$$C(k) = \phi k^2, \tag{7}$$

where ϕ is the investment coefficient, which is not changing for developing quality. To compete with the increasing market demand of the product, it is required to increase the quality of goods.

Cost for buying used products

The distributor (or the 3PL) gives the remainder portion of used products $(1 - x)$ to the manufacturing company, then the manufacturing company pays b per unit used product to the distributor (or the 3PL). For buying a used product, there is a cost known as product return cost.

$$rL = b(1 - x)R(j, k) \tag{8}$$

Remanufacturing cost

The manufacturer remanufacturers $(1 - x)$ portion of used products. For remanufacturing, the returned product requires c_1 unit remanufacturing cost per product. The remanufacturing cost is

$$rF = c_1(1 - x)R(j, k)). \tag{9}$$

Within this representation, manufacturer only collects end-of-life products for remanufacturing from a distributor or 3PL.

Manufacturing cost of new product

Within this model, each unit new product manufacturing cost is c_m. Thus, the total manufacturing cost of the new product is dependent on the order quantity (M) of the product. The manufacturer produces $(M - R(j, k))$ amount of new products, where $R(j, k)$ is the remanufactured used products. The manufacturing cost is

$$NPC = C_m(M - R(j, k)). \tag{10}$$

Each new product's cost is fixed.

Carbon tax

The manufacturer bears carbon emission costs under a carbon cap and with trade strategy. The manufacturer invests CO_2 emission tax to decrease CO_2 emission for the green environment and benefit of the supply chain. The manufacturer has the emission cap of G_v grams from the government. If the total emissions for producing M quantity of products exceeds the carbon cap, then the manufacturer pays carbon tax for that extra carbon emission. Besides, the manufacturer emits carbon in a reduced rate of v. The total carbon emission cost is

$$Cec = P_v[\eta(1 - v)M - G_v]. \tag{11}$$

This cost has a definite effect on reducing CO_2 emission. In this scenario, the manufacturer pays the carbon tax to the government for recycling the products.

Revenue

The manufacturer generates revenue by selling products to the distributor at a wholesale price w. The wholesale price for new product is $w(M - R(j, k))$. Wholesale price and cost savings from the remanufactured products is $(w + T_1)(1 - x)R(j, k)$. Besides, the manufacturer receives price for technology sharing as $lxR(j, k)$. Thus, total revenue of the manufacturer is $w(M - R(j, k)) + (w + T_1)(1 - x)R(j, k) + lxR(j, k)$.

4.2 Distributor's Mathematical Model

The distributor takes major roles in the supply chain for product selling and maximizing profit in the supply chain. The distributor always gives holding costs and shortage costs. In one scenario, the distributor remanufactures the used products as new by his technical ability and sells them in the market to earn a profit along with new products (Scenario E). In the other scenario (Scenario F), the distributor earns a profit margin by selling new and remanufactured reproducts but does not take part in remanufacturing. To increase the demand for products in the market, the distributor uses a servicing investment. Here, distributor's optimal profit is calculated by subtracting all costs from the revenue.

Again $E(M)$ is the anticipated trading amounts, which is acquired by using the following equation:

$$\begin{aligned} E(M) &= \int_0^M x(D)D\mathrm{d}D + [1 - X(M)]M \\ &= M - \int_0^M X(D)\mathrm{d}D. \end{aligned} \tag{12}$$

Selling price

The distributor sells products in the market and generates revenue. If M be the total quantity to sell and w be the product's unit selling price, then the total selling price is wM.

Holding cost

s^+ is the unit holding cost of products that are not sold in the market. $Z(M)$ is the function of the distributor's inventory leftover.

$$\begin{aligned} Z(M) &= (M + R(j,k)x - D)^+ \\ &= R(j,k)x + M - E(M) \end{aligned} \tag{13}$$

Thus, total holding cost is $s^+Z(M)$.

Shortage cost

Product shortage is a negative effect on the market. The shortage cost is s^- per unit product. $S(M)$ is the function of the distributor's expected lost sales.

$$\begin{aligned} S(M) &= (D - R(j,k)x - M)^+ \\ &= E(D) - R(j,k)x - E(M) \end{aligned} \tag{14}$$

Thus, total shortage cost is $s^{-}S(M)$.

Cost for buying used product

The return product quantity and return product cost are proportional, but the supply chain's return product cost and profit are inversely proportional. In this model, the distributor pays the compensated price j per unit return product. If the quantity of the returned product is $R(k, j)$, then the total cost for returned products is $jR(k, j)$.

Remanufacturing cost

The distributor is remanufacturing the quantity of x portion of the returned products. Thus, the total cost for remanufacturing is

$$rD = c_2 R(j, k)x. \tag{15}$$

In this model, remanufacturing cost and return product quantity are in a proportional relationship.

Cost for technology licensing fees

When the distributor takes part in the remanufacturing process, then the distributor needs a technology license for remanufacturing. The distributor buys a technology license from the manufacturer. The technology licensing fees per unit item is l, then total technology licensing fees for remanufacturing is

$$Tc = R(j, k)xl. \tag{16}$$

The manufacturer gives the authority to the distributor of supply chain to remanufacturer the returned products that are similar quality as the manufacturers.

Servicing investment

The distributor bares a servicing investment for increasing the market demand. This service gains customer's attraction and increases the selling quantity. The servicing investment for the products is below.

$$\text{sic} = \alpha \frac{s^2}{2} \tag{17}$$

4.3 Mathematical Model of Third Party Logistics (3PL)

The 3PL is a collector of used products from customers by giving them refund price. A 3PL comes into the CLSCM when the distributor does not take part in the remanufacturing process. Then, the manufacturer shares the technology license with the 3PL for remanufacturing (Scenario F). In this case, the 3PL only remanufacturers a portion of used products. The distributor sells all products to the market. In this case, the 3PL's optimal profit is calculated by subtracting all the costs from revenue earned by the 3PL.

Remanufacturing cost

Here x portion of the total returned products $R(k, j)$ is remanufactured by the 3PL each unit cost for remanufacturing is c_3. Thus, the total cost for remanufacturing is

$$Rt = c_3 x R(k, j). \tag{18}$$

In this scenario, 3PL remanufactures the return products based on their remanufacturing ability. Here, we consider a remanufacturing process by a manufacturer and third party.

Cost for technology licensing fees

The 3PL pays a per unit licensing cost to the manufacturer for buying the technology license. In this case, l is the licensing fees for remanufacturing each unit product. The entire technology licensing fees cost is

$$TL = l x R(k, j). \tag{19}$$

In this model, the 3PL considers himself partners for remanufacturing the products under the technology license taken from the manufacturer to decrease manufacturing costs.

Cost for buying used item

In this context, the 3PL pays customers an exchange price b for per unit end-of-life used product. For x unit returned products, the cost of a 3PL is

$$Rct = b x R(k, j) \tag{20}$$

In this scenario, a higher return product cost positively affects return product quantity but negatively affects supply chain profit.

Now we consider two scenarios E and F for profit of the supply chain which are as follows.

4.3.1 Scenario *E*

End-of-life used products are remanufactured by the manufacturer and distributor under a technology licensing sharing contract. The distributor remanufacturers a portion of used products and sells products in market. 3PL is not a part of this CLSCM. Scenario E forms a two-echelon CLSCM.

By the previous discussion, we get the profit function of the manufacturer in Scenario E.

$$\begin{aligned}\pi_F^E(k, v) &= \text{Revenue} - c(v) - c(k) - rL - rM - \text{NPC} - \text{cec} \\ &= (w - C_M)(M - R(j, k)) - P_v[\eta(1 - v)M - G_v] \\ &\quad - \phi k^2 + (w - C_M + T_1 - b)(1 - x)R(j, k) \\ &\quad + lxR(j, k) - \theta\frac{v^2}{2}\end{aligned} \tag{21}$$

The distributor's expected profit is determined as

$$\begin{aligned}E(\pi_Y) &= \text{Revenue} - \text{PC} - \text{HC} - \text{SC} - \text{RPC} - \text{RC} - \text{TC} - \text{sic} \\ &= pE(M) - wM - s^+Z(M) - s^-S(M) \\ &\quad - jR(j, k) + R(j, k)x(w - C_M + T_2 - l) \\ &\quad + b(1 - x)R(j, k) - \alpha\frac{s^2}{2}\end{aligned} \tag{22}$$

where,

$$\begin{aligned}M &= I + a_0 + a_1\frac{p_{\max} - p}{p - p_{\min}} + a_2 v + a_3 v^2 + a_4 j + a_5 j^2 \\ &\quad + a_6 k + a_7 k^2 + a s^{\gamma}.\end{aligned} \tag{23}$$

For calculation, with originality, the components of the distributor's profit are transferred from the ordering quantity (M) to the expected inventory (I). Keeping this in mind, Eq. (24) must hold.

$$\int_0^M x(D)\mathrm{d}D = \int_0^I x(\epsilon)\mathrm{d}\epsilon \tag{24}$$

Accordingly, the estimated profit function of distributor is represented as

$$\begin{aligned}\pi_Y^E(j, I, s, p) &= (p - w)(M) + s^-(I - \rho) - (p + s^+ \\ &\quad + s^-)\int_0^I X(\epsilon)\mathrm{d}\epsilon + [(w - c_M + T_2 - l \\ &\quad - b + s^- - s^+)x + b - j](\mu + cj - yk) - \alpha\frac{s^2}{2}\end{aligned} \tag{25}$$

The diagram of scenario E is shown in Fig. 1.

4.3.2 Scenario *F*

In this scenario, the manufacturer shares technology license with the 3PL. The products are remanufactured by the manufacturer and 3PL in a hybrid remanufacturing. Then, this forms a three-echelon CLSCM, where the distributor only sells products to the market.

The manufacturer profit function for Scenario F is

$$\begin{aligned}\pi_F^F(v,k) &= \text{Revenue} - c(v) - c(k) - rL - rM - \text{NPC} - \text{Cec}\\ &= (w - C_M)(M - R(k,j)) - P_v[\eta(1-v)M - G_v] - \theta(\frac{v^2}{2})\\ &\quad - \phi k^2 + (w - C_M + T_1 - b)(1-x)R(k,j) + lxR(k,j) \qquad (26)\end{aligned}$$

which simplifies to

$$\begin{aligned}\pi_F^F(v,k) = (w - C_M - P_v\eta + P_v\eta v)M + P_vG_v - \theta(\frac{v^2}{2}) - \phi k^2\\ + R(k,j)[(l + C_M - w)x + (T_1 - b)(1-x)]. \qquad (27)\end{aligned}$$

Furthermore, the expected profit of the distributor is as below

$$\begin{aligned}\pi_Y^F(I,s,p) &= \text{Revenue} - \text{PC} - \text{HC} - \text{SC} - \text{sic}\\ &= (p-w)M + s^-(I-\rho) - (s^- + p + s^+)\int_0^I X(\epsilon)\text{d}\epsilon - \alpha\frac{s^2}{2} \qquad (28)\end{aligned}$$

Besides, the profit function of the 3PL is

$$\begin{aligned}\pi_H^F(j) &= \text{Revenue} - \text{RC} - \text{TL} - \text{RCT}\\ &= [(w - C_M + T_3 - b - l)x + b - j](\mu + cj - yk). \qquad (29)\end{aligned}$$

The diagram of scenario F is shown in Fig. 1.

5 Solution Methodology

Within the context, the main target is to optimize decision variables v, k, s, I, j, and p for optimal profit of the CLSCM in two scenarios E and F, respectively.

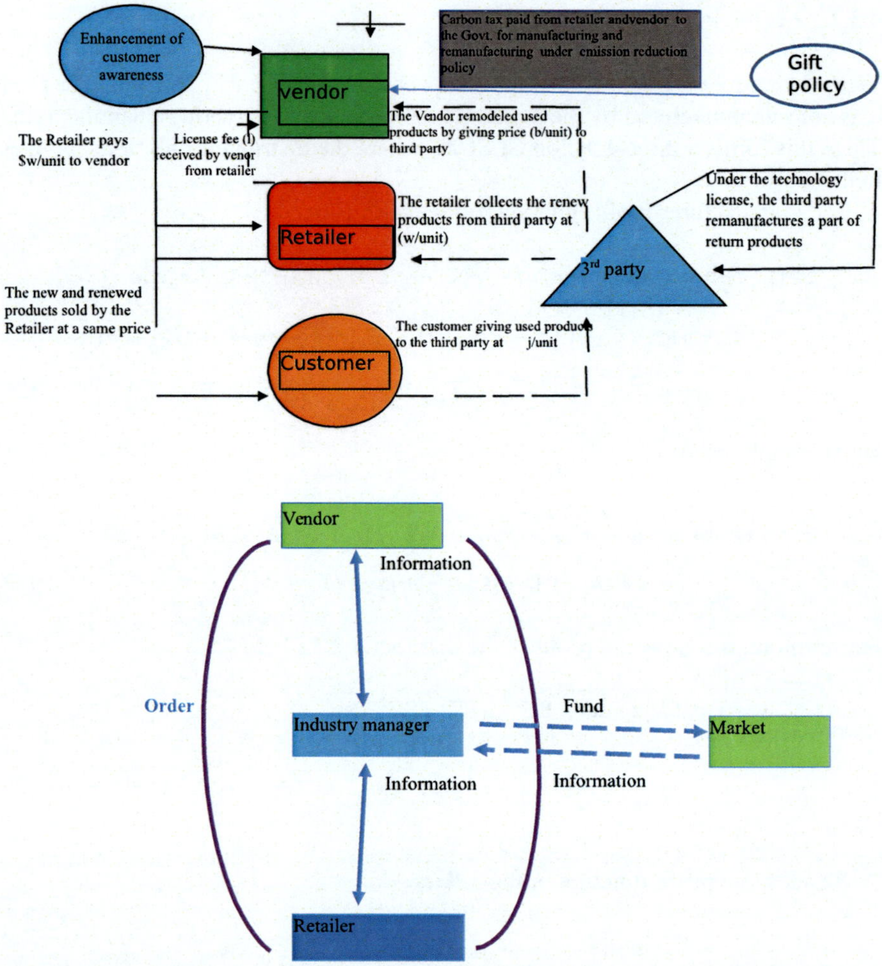

Fig. 1 The proposed hybrid remanufacturing system within a three-echelon CLSCM

5.1 Stackelberg Game in Scenario E

In this case, manufacturers acts as a leader and the distributor follows him. Using optimal I, j, p, s in manufacturer's profit function and we get optimal v, k. The values of decision variables of the distributor is mentioned below.

Proposition 1 In this case, the equilibrium condition is applied to the distributor's profit to achieve the distributor's optimum decision variables I^* , j^*, s^*, p^* in Eqs. (30)–(33) which are given below.

$$I^* = \frac{k_1(p^* - w + s^-)}{p^* + s^+ + s^-} \tag{30}$$

$$j^* = \frac{X_1}{-2a_5(p^* - w) + 2c} \tag{31}$$

where
$X_1 = -\mu + yk^* + c(w - c_M + T_2 - l - b)x - bc - (p^* - w)a_4$

$$s^* = [\frac{\alpha}{(p^* - w)a\gamma}]^{\frac{1}{\gamma-2}} \tag{32}$$

$$p^* = \frac{-2p_{\min}(a_1-L_1+L_2)+\sqrt{((a_1-L_1+L_2)2p_{\min})^2-4L_5(L_1-L_2-a_1)}}{2(L_1-L_2-a_1)} \tag{33}$$

where , $H_1 = I + a_0 + a_2 v^* + a_3 v^{*2} + a_4 j^* + a_5 j^{*2} + a_6 k^* + a_7 k^{*2} + as^{*\gamma} + \frac{1}{2k_1}$ and $H_2 = p_{\max} H_1 + a_1 w(p_{\max} - p_{\min}) + p_{\min} p_{\max}$.

Later, to find out the manufacturer optimal decision variables v^*, k^*, we apply equilibrium condition on the manufacturer's profit function.

Proposition 2 The manufacturer's decision variables v^*, k^* are define in Eqs. (34) and (35) as below:

$$v^* = \frac{-H_4 + \sqrt{H_4^2 - 12a_3 H_6 p_v \eta}}{6a_3 p_v \eta}. \tag{34}$$

$H_3 = I + a_0 + a_1 \frac{p_{\max} - p}{p - p_{\min}} + a_4 j + a_5 j^2 + a_6 k + a_7 k^2 + as^{\gamma}$,
$H_4 = 2a_3(w - p_v \eta - c_M) + 2a_2 p_v \eta - \theta$,
and $H_6 = H_3 p_v \eta + a_2(w - c_M - p_v \eta)$

$$k^* = \frac{y(-b+x(b+l-w+c_M-T_1)+T_1)-a_6(w-c_M-p_v\eta+p_v v\eta)}{2a_7(w-c_M-p_v\eta+p_v v\eta)-2\phi} \tag{35}$$

5.2 *Stackelberg Game in Scenario F*

In this case, the manufacturer acts as a leader, then the distributor, and 3PL follow the manufacturer. Here putting the optimal values of I, j, p, s in the manufacturer profit function, we get v, k .

Proposition 3 In this case, applying equilibrium condition on 3PL and distributor's profit function, the optimal decision variables j^*, I^*, s^*, p^* are given in Eqs. (36)–(39).

$$j^* = \frac{yk^* - \mu + bc + cx(T_3 + w - c_M - b - l)}{2c} \tag{36}$$

$$I^* = \frac{k_1(p^* - w + s^-)}{s^- + p^* + s^+} \tag{37}$$

$$s^* = [\frac{\alpha}{(p^* - w)a\gamma}]^{\frac{1}{\gamma - 2}} \tag{38}$$

$$p^* = \frac{-2p_{\min}(a_1 - L_1 + L_2) + \sqrt{((a_1 - L_1 + L_2)2p_{\min})^2 - 4L_5(L_1 - L_2 - a_1)}}{2(L_1 - L_2 - a_1)}. \tag{39}$$

where, $L_1 = I + a_0 + a_2 v^* + a_3 v^{*2} + a_4 j^* + a_5 j^{*2} + a_6 k^* + a_7 k^{*2} + as^{*\gamma}$ and $L_2 = \frac{I^2}{2k_1}$.
Also, $L_5 = p_{\min}^2(L_1 - L_2) - a_1 p_{\max} p_{\min} + a_1(w(p_{\max} - p_{\min}))$.

To find out the manufacturer's optimal decision variables v^*, k^*, we are applying equilibrium condition on the manufacturer's profit function.

Proposition 4 In this case, applying equilibrium condition on the manufacturer profit function, optimal decision variables v^*, k^* are defined in Eqs. (40) and (41).

$$v^* = \frac{-L_4 + \sqrt{L_4^2 - 12a_3 L_6 p_v \eta}}{6a_3 p_v \eta} \tag{40}$$

where , $L_3 = I + a_0 + a_1 \frac{p_{\max} - p}{p - p_{\min}} + a_4 j + a_5 j^2 + a_6 k + a_7 k^2 + as^\gamma$,
and , $L_4 = 2a_3(w - p_v \eta - c_M) + 2a_2 p_v \eta - \theta$.
Also, $L_6 = L_3 p_v \eta + a_2(w - c_M - p_v \eta)$

$$k^* = \frac{y(-b + x(b + l - w + c_M - T_1) + T_1) - a_6(w - c_M - p_v \eta + p_v v \eta)}{2a_7(w - c_M - p_v \eta + p_v v \eta) - 2\phi} \tag{41}$$

6 Numerical Analysis

This study uses input data from Mondal et al. [14] and Taleizadeh et al. [25]. The proposed analytical solution process is used to solve numerical examples.

6.1 Example 1

Example 1 gives numerical results for Scenario E. Table 2 gives input value of parameters. Total profit and optimum values of decision variables for Scenario E are shown in Table 3.

Table 2 Parametric values for Scenario E and F

Parameters	Value	Parameters	Value
$[k_2, K_1]$	[0,90]	w	$70/unit
c_1	$5/unit	c_3	$5/unit
P_v	$0.2/gallons	G_v	100 gallons
η	2.99	c	1.22
μ	30 units	y	0.6
p_{max}	$200/unit	p_{min}	$90/unit
γ	3.65	a	30
a_0	2000 Units	a_1	0.7
a_2	0.2	a_3	0.2
a_4	0.5	a_5	0.4
a_6	5	a_7	0.3
ϕ	$1000	α	$1000
x	0.4	l	$50/unit
B	40	ρ	0.2

Table 3 Optimal solutions and maximum profit of Scenario E

I units	s	$\$j$/unit	$\$p$/unit	v
17.75	0.52	5.73	80.10	0.56
K				
0.11				
$\$\pi_r^U$	$\$\pi_t^U$	$\$\pi_m^U$	Total profit ($)	
687.15	41,161.94	96,309.23	138,158.32	

Table 4 Optimal solutions and maximum profit of Scenario F

I units	s	$\$j$/unit	$\$p$/unit	v
5.96	0.76	7.87	88	0.58
K				
0.12				
$\$\pi_r^{DF}$	$\$\pi_t^{DF}$	$\$\pi_m^{DF}$	**Total profit ($)**	
187.15	40,789.35	95,209.09	136,185.60	

6.2 Example 2

Example 2 gives numerical results for Scenario F. Table 2 gives the parametric value for Example 2. Maximum profit and optimal values of decision variables are given in Table 4.

Results show that Scenario E has more profit than Scenario F. This happens because Scenario E has a known distribution function. Thus, all information about demand is known to the management and thus, it is easy to optimize the objective. But,

Table 5 The sensitivity investigation of important cost parameters of this study

Parameters	Changes(%)	3PL's profit changes(%)	Distributor's profit changes(%)	Manufacturer's profit changes(%)
w	−50	−49.45	+108.58	−78.12
	−25	−26.31	+53.13	−47.39
	+25	+29.39	−81.91	+77.79
	+50	+61.74	−234.54	+199.89
b	−50	−13.47	−10.52	−10.55
	−25	−6.84	−5.62	−5.64
	+25	+7.04	+6.34	+6.37
	+50	+14.28	+13.41	+13.47
c_m	−50	+16.31	+16.77	+40.55
	−25	+8.05	+7.84	+18.82
	+25	−7.81	−6.77	−16.25
	+50	−15.38	−12.48	−30.29
p_v	−50	−0.008	+0.013	+0.60
	−25	−0.002	+0.004	+0.22
	+25	−0.0008	+0.004	+0.07
	+50	−0.008	+0.02	+0.59
h	−50	−4.93		
	−25	+5.48		
	+25	+5.48		
	+50	+5.48		

in reality, there are a lot of risks and uncertainties that evolve along with information. Then, the management needs to justify that information more than the known case, as Scenario E.

6.3 Sensitivity Analysis

The sensitivity investigation of the main specifications of this study is summarized in Table 5 based on the Example 1.

7 Managerial Insights

Some managerial insights derived from this study are given below.

1. The industry manager must focus on the hybrid remanufacturing process where the 3PL and the manufacturer both remanufacture the return products (scenario F), and distributor only sells the products. It is lesser successful policy to earn more profit than Scenario E. Therefore, the manufacturer with information about the best remanufacturing strategy can reach the environmental goal (Figs. 2–3).

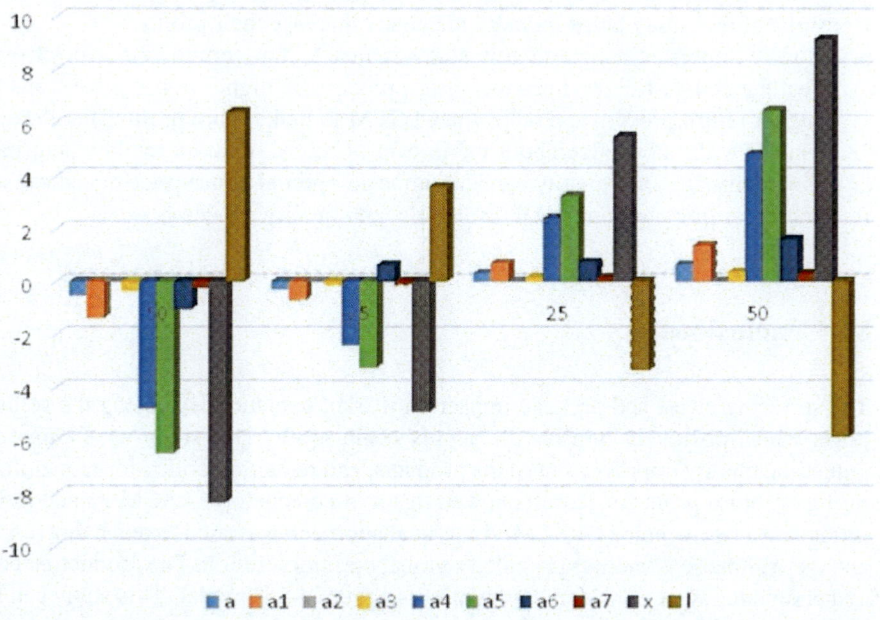

Fig. 2 Sensitivity analysis of cost parameters

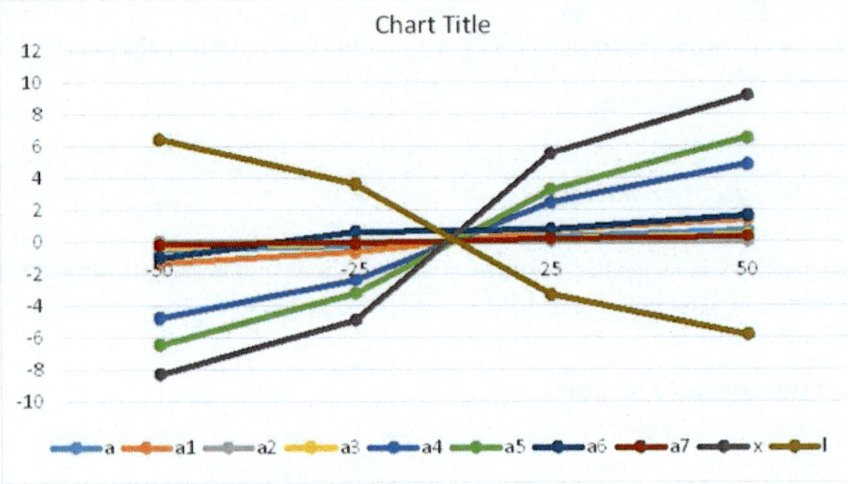

Fig. 3 Sensitivity analysis of cost parameters

2. Managers of the industry must focus on servicing efforts. This paper analyzes the impact of servicing actions on total profits. In order to increase the demand for the product, some costs should be invested for the purpose of service to customers. In this competitive market, this service strategy plays an important role. Thus, the results of this study helps industry managers increase their profits.
3. Another important aspect of this study is that higher refund prices positively affect the end-of-life used product return process. A higher return price leads to a higher return process and helps the CLSCM to make more profit. Therefore, a manager with knowledge about the impact of return price on remanufacturing, carbon emission, and quality can determine an optimal return price to achieve all goals regarding environmental, economic, and quality concerns.

8 Conclusions

The servicing effort had an huge impact on market demand, decreasing the selling price, which positively impacts the supply chain profit. This study used emission reduction, quality improvement of the products, and reduction policy for remanufacturing of return products. Result showed that a three-echelon CLSCM gained more profit than a two-echelon CLSCM. Analytical approach was used to reach this result, and the arithmetical findings provided a global optimal solution. The production cost and carbon tax were considered as continuous type in this model. This study can be extended by using discrete type of price and tax. Besides, some other motivation factors like customer awareness are not considered in this research. In the future model, these factors can be considered.

References

1. Allan, G., Lecca, P., McGregor, P., Swales, K.: The economic and environmental impact of a carbon tax for Scotland: a computable general equilibrium analysis. Ecol. Econ. **100**, 40–50 (2014)
2. Kar, S., Basu, K., Sarkar, B.: Advertisement policy for dual-channel within emissions-controlled flexible production system. J. Retail. Consum. Serv. **71**, 103077 (2023)
3. Chen, J.M., Chang, C.I.: The co-opetitive strategy of a closed loop supply chain with remanufacturing. Transp. Res. Part E Logistics Trans. Rev. **48**(2), 387–400 (2012)
4. Datta, T.K.: Effect of green technology investment on a production inventory system with carbon tax. Adv. Oper. Res. (2017)
5. Demirel, E., Demirel, N., Cen, H.G.: A mixed integer linear programming model to optimize reverse logistics activities of end-of-life vehicles in Turkey. J. Cleaner Prod. **112**(3), 2101–2113 (2016)
6. Khan, I., Malik, A.I., Sarkar, B.: A distribution-free newsvendor model considering environmental impact and shortages with price-dependent stochastic demand. Math. Biosci. Eng. **20**(2), 2459–2481 (2023)
7. Giovanni, P.D.: A joint maximization incentive in closed-loop supply chains with competing retailers: the case of spent-battery recycling. Eur. J. Oper. Res. **268**(1), 128–147 (2018)
8. Giri, B.C., Sharma, S.: Optimal production policy for a closed loop hybrid system with uncertain demand and return under supply disruption. J. Cleaner Prod. **112**(3), 2015–2028 (2016)
9. Harris, L.C., Ogbonna, E.: Exploring service sabotage: the antecedents, types and consequences of frontline, deviant, antiservice behaviors. J. Serv. Res. **4**(3) (2002)
10. Jacobs, B.W., Subramanian, R.: Sharing responsibility for product recovery across the supply chain. Prod. Manage. **21**(1), 85–100 (2012)
11. Ji, J., Zhang, Z., Yang, L.: Carbon emission reduction decisions in the retail-/dual-channel supply chain with consumers' preference. J. Cleaner Prod. **141**, 852–867 (2017)
12. Liu, Y., Holzer, J.T., Ferris, M.C.: Extending the bidding format to promote demand response. Energy Policy **86**, 82–92 (2015)
13. Liu, L., Huang, C.Z., Huanga, G., Baetzd, B., Pittendrigha, S.M.: How a carbon tax will affect an emission-intensive economy: a case study of the Province of Saskatchewan. Canada. Energy **159**, 817–826 (2018)
14. Mondal, A.K., Pareek, S., Chaudhuri, K., Bera, A., Bachar, R.K., Sarkar, B.,: Technology license sharing strategy for remanufacturing industries under a closed-loop supply chain management bonding. RAIRO-Operat Res. **56**(4), 3017–3045 (2022)
15. Qin, J., Bai, X., Xia, L.: Sustainable trade credit and replenishment policies under the cap-and-trade and carbon tax regulations. MDPI, Open Access J. **7**(12), 16340–16361 (2015)
16. Ramani, V., Giovanni, D.P.: A two-period model of product cannibalization in an atypical closed-loop supply chain with endogenous returns: the case of DellReconnect. Eur. J. Oper. Res. 1009–1027 (2017)
17. Chaudhari, U., Bhadoriya, A., Jani, M.Y., Sarkar, B.: A generalized payment policy for deteriorating items when demand depends on price, stock, and advertisement under carbon tax regulations. Math. Comp. Simul. **207**, 556–474 (2023)
18. Kugele, A.S.H., Sarkar, B.: Reducing carbon emissions of a multi-stage smart production for biofuel towards sustainable development. Alexan. Eng. J. **70,** 93–113 (2023)
19. Yadav, D., Chand, U., Goel, R., Sarkar, B.: Smart production system with random imperfect process, partial backordering, and deterioration in an inflationary environment. Mathematics **11**(2), 440 (2023)
20. Savaskan, R.C., Wassenhove, L.V.: Reverse channel design: the case of competing retailers. Manage. Sci. **52**(1), 1–14 (2006)
21. Savaskan, K.C., Bhattacharya, S., Wassenhove, V.: Closed-loop supply chain models with product remanufacturing. Research Collection Lee Kong CHIAN School of Business, Publication Type, Journal Article (2004)

22. Shi, J., Zhanga, G., Shab, J.: Optimal production planning for a multi-product closed loop system with uncertain demand and return. Comput. Oper. Res. **38**(3), 641–650 (2011)
23. Taleizadeh, A., Zerang, E.S., Choi, T.M.: The effect of marketing effort on dual-channel closed-loop supply chain systems. Systems 1–12 (2016)
24. Taleizadeh, A., Moshtagh, M.S., Moon, I.: Optimal decisions of price, quality, effort level and return policy in a three-level closed-loop supply chain based on different game theory approaches. Eur. J. Ind. Eng. **11**(4), 486 (2017)
25. Taleizadeh, A.A., Basban, N.A., Niaki, S.T.A.: A closed-loop supply chain considering carbon reduction, quality improvement effort, and return policy under two remanufacturing scenarios. J. Cleaner Prod. **232**, 1230–1250 (2019)
26. Teunter, R., van der Laan, E.: On the non-optimality of the average cost approach for inventory models with remanufacturing. Int. J. Prod. Econ. **79**(1), 67–73 (2002)
27. Wang, Y., Chena, W., Liu, B.: Manufacturing/remanufacturing decisions for a capital-constrained manufacturer considering carbon emission cap and trade. J. Cleaner Prod. **140**(3), 1118–1128 (2017)
28. Xu, M., Qi, X., Yu, G., Zhang, H., Gao, C.: The demand disruption management problem for a supply chain system with nonlinear demand functions. J. Syst. Sci. Syst. Eng. **12**, 82–97 (2003)
29. Xu, L., Li, Y., Govindan, K., Yue, X.: Return policy and supply chain coordination with network-externality effect. Int. J. Prod. Res. **56**(10), 3714–3732 (2018)
30. Yang, G., Wang, Z., Li, X.: The optimization of the closed-loop supply chain network. Transp. Res. Part E Logistics Transp. Rev. **45**(1), 16–28 (2009)
31. Zeng, J., Yao, Q., Wang, X., Zhang, Y.: Game research on coal mine workers' off-post behaviors. Math. Probl. Eng. (2019)
32. Zhao, R., Liu, Y., Zhang, N., Huang, T.: An optimization model for green supply chain management by using a big data analytic approach. J. Cleaner Prod. **142**(2), 1085–1097 (2017)

A Review on Type-2 Fuzzy Systems in Robotics and Prospects for Type-3 Fuzzy

Fevrier Valdez, Oscar Castillo, and Patricia Melin

Abstract In this article, we are offering a review of type-2 fuzzy systems and their current applications in robotics. Although, significant research work has been put forward in to showing that fuzzy systems have very good capabilities for coping with uncertainty in control applications, there is still room for improvement and this is why the interest of doing this research work. Nowadays, fuzzy logic systems are used frequently to manage uncertainty because the obtained results have been superior to traditional methods. However, when the uncertainty on the problems is high fuzzy logic is not able of manage adequately the uncertainty. For that reason, the authors are utilizing type-2 fuzzy logic to obtain better results on the control problems. In this article, we have made a review over the papers using type-1 and type-2 fuzzy systems, specifically when they are utilized in robotics. The analysis was made with several searches, for example, using nature optimization methods with type-2 fuzzy in robotics and also without the utilization of optimization methods. The collection of papers was obtained from Web of Science (WoS) and the visual tool 'connected papers'. We also briefly discuss the prospects for the utilization in the future of type-3 fuzzy systems in robotics.

Keywords Type-2 fuzzy logic · Robotics · Optimization · Type-3 fuzzy systems

1 Introduction

The main motivation of this article was performing an exhaustive review of fuzzy systems and their applications in different kinds of robotic systems. In addition,

F. Valdez · O. Castillo (✉) · P. Melin
Tijuana Institute of Technology, TecNM, Calzada Tecnologico S/N, 22450 Tijuana, Mexico
e-mail: ocastillo@tectijuana.mx

F. Valdez
e-mail: fevrier@tectijuana.mx

P. Melin
e-mail: pmelin@tectijuana.mx

O. Castillo et al. (eds.), *Applied Mathematics and Computational Intelligence*,
Springer Proceedings in Mathematics & Statistics 413,
https://doi.org/10.1007/978-981-19-8194-4_17

as a natural evolution of fuzzy theory, we also discuss the possible utilization of type-3 fuzzy logic in robotics and control. Nowadays, it is very popular the use of type-2 techniques [31] to improve the performance in several areas, for instance, in [19] it was proposed a type-2 fuzzy logic controller (FLC) for the real-time control of mobile robots [25, 28, 37, 12]. The authors used the type-2 FLC to design and implement a variety of robotic behaviors on different platforms in indoor and outdoor unstructured and challenging dynamic (changing) environmental situations. According to site https://www.connectedpapers.com/ and taking as reference the work above mentioned, in Fig. 1, we found the connected papers with the reference [19], as can be seen, since 2004 the number of works in the area has been increasing steadily.

Also, in Table 1, the derivative works of the paper [19] are presented.

These are papers that cited many of the papers in the graph. This means that they are either surveys of the field or recent relevant works which were inspired by many papers in Figure 1.

Also, in [8], A New Meta-Heuristics of Optimization with Dynamic Adaptation of Parameters using Type-2 Fuzzy Logic for Trajectory Control of a Mobile Robot was proposed. Also, according to site https://www.connectedpapers.com/ and taking as reference the paper [8], in Figure 2, we found the connected papers with the reference [8], as can be seen, this work is more recent than the before analyzed work. However, the number of works is increasing in the last years. Also, Table 2 is presented the derivative works of the paper [8]. These are papers that cited many of the papers

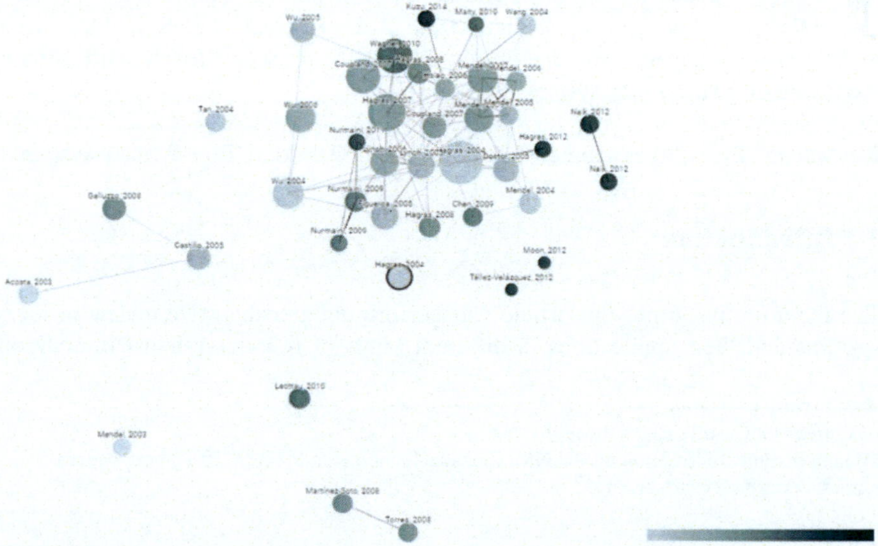

Fig. 1 Connected papers with the paper 'A type-2 fuzzy logic controller for autonomous mobile robots' [19]

Table 1 Derivative works of the paper by Hagras in [19]

Paper	Last author	Year	Citations
Advances in type-2 fuzzy sets and systems [30]	Jerry M., Mendel	2007	344
Derivation and analysis of the analytical structures of the interval type-2 fuzzy-PI and PD controller [14]	Hao, Ying	2010	77
Adaptive type-2 fuzzy logic control of a bioreactor [18]	Bartolomeo, Cosenza	2010	9
A dynamic defuzzification method for interval type-2 fuzzy logic controllers [48]	Ibrahim, Eksin	2011	6
Systematic design of type-2 fuzzy logic systems for modeling and control with applications to modular and reconfigurable robots [6]	Mohammad, Biglarbegian	2010	2
Analysis and control of mobile robots in various environmental conditions [18]	S., Mahapatra	2012	1
Control of an ambiguous real time system using interval type 2 fuzzy logic control [50]	Deepa, Thangavelusamy	2017	1
A comparison of type-1 and Type-2 fuzzy logic controllers in robotics: a review [39]	Syed Wakil, Ah-mad	2015	0
Circumventing the fuzzy type reduction for autonomous vehicle controller [1]	Al-Rikabi	2017	0
Novel fuzzy techniques for modelling human decision making [33]	Salang, Musika	2013	0

in the graph. This means that they are either surveys of the field or recent relevant works which were inspired by many papers in Fig. 2.

The remaining sections of this article are formed in the following fashion. In Section 2, a sample of papers from the review of the literature is offered. In Section 3, the review of Type-2 fuzzy logic in robotic applications is shown, and Section 4 outlines the review with Type-1 fuzzy logic in robotics. Lastly, in Section 5, the conclusions are outlined.

2 Literature Review

In this section, we briefly describe a sample of the papers that have utilized type-2 fuzzy in robotic applications. These papers offer an idea of the kind of works that have been done in this area.

In [36], controlling a robotic manipulator was intended under significant external perturbations and parametric uncertainties. The authors decided to use Type-2 fuzzy logic as an adequate choice to cope with the uncertainty of dynamic environments,

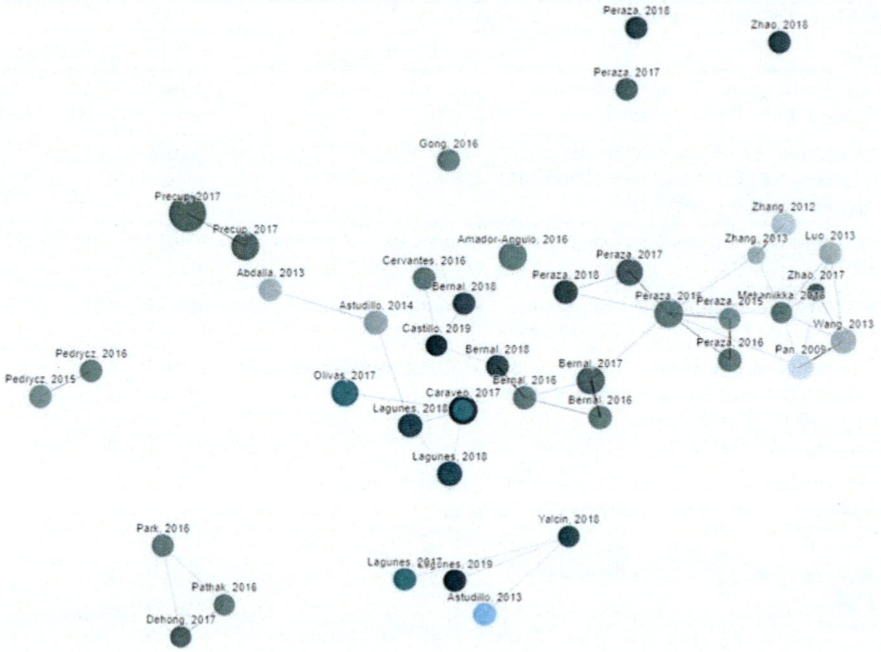

Fig. 2 Connected papers with the paper 'a new meta-Heuristics of optimization with dynamic adaptation of parameters using type-2 fuzzy logic for Trajectory control of a mobile robot' [8]

for example, the utilization of membership functions with fuzzy values. Also, they used a neural network to achieve additional robustness behavior. In this case, the initial rules are based on a sliding surface of a controller, and in this way, it can improve the system's performance.

In [44], a review of recent type-2 fuzzy controller applications was described. The authors made a survey about the applications of controllers, encompassing areas such as robotics, manufacturing systems, electrical systems, and aircraft control. The most promissory ones have been found to be in the robotics and automotive areas, where type-2 FLCs have exhibited to perform better than their type-1 fuzzy counterparts and also with respect to traditional controllers.

In [40] an alternative systematic methodology to explicitly derive membership functions for both type-1 FLC and interval type-2 FLC was presented. The proposed approach to obtain a closed-form relationship between inputs and output offers information on their impact on the footprint of uncertainty parameters. Also, in [23], a novel application of genetic algorithms (GA) optimization approach to optimize the scaling factors of interval type-2 fuzzy proportional derivative plus integral controllers is proposed for 5-degrees of freedom redundant robot manipulator for trajectory tracking task.

Table 2 Derivative works of the paper [8]

Paper	Last author	Year	Citations
Optimization of membership function parameters for FLCs of an autonomous mobile robot using the flower pollination algorithm [38]	José, Soria	2018	3
Path Finding for a Mobile Robot Using Fuzzy and Genetic Algorithms [26]	Arbnor, Pajaziti	2017	2
Comparative Study in FLC Optimization using Bee Colony, Differential Evolution, and Harmony Search Algorithms [9]	Cinthia, Peraza	2019	2
Optimization of FLCs Using Galactic Swarm Optimization with Type-2Fuzzy Dynamic Parameter Adjustment [4]	Fevrier, Valdez	2019	2
Comparative Study of the Conventional Mathematical and FLC for Velocity Regulation [49]	Cinthia, Peraza	2019	1
Shadowed Type-2 Fuzzy Systems for Dynamic Parameter Adaptation in Harmony Search and Differential Evolution Algorithms (Castillo et al. 2019b)	Patricia, Ochoa	2019	1
Intelligent Information and Database Systems [35]	Bogdan, Trawin ski	2019	0
Interval Type-2 fuzzy logic for dynamic parameter adjustment in the imperialist competitive algorithm [5]	Fevrier, Valdez	2019	0
On Characterizations of Directional Derivatives and Sub differentials of [51]	Dong, Qiu	2017	0

In [11], a type-2 fuzzy logic controller (FLC) is proposed, in their work for robot manipulators with joint elasticity and structured and unstructured dynamical uncertainties, a controller is based on a sliding mode control strategy. To enhance the real-time functioning of the controller, the simpler interval type-2 fuzzy sets were used. On the other hand, in [22], a novel concept of an interval type-2 fractional-order fuzzy PID (IT2FO-FPID) controller, which requires fractional-order integrator and fractional-order differentiator, is proposed. The incorporation of Takagi–Sugeno-Kang (TSK) type interval type-2 fuzzy logic controller (IT2FLC) with fractional controller of PID-type is investigated for time response measure due to both unit step response and unit load disturbance. Also, in [16], a computationally efficient systematic procedure to design an optimal type-2 fuzzy logic controller was proposed. The most important idea was to optimally find the controller gains using the particle swarm algorithm, and then optimally find the membership function parameters utilizing a more basic genetic algorithm.

In [32], an H-infinity output feedback controller is developed for a class of time-delayed MIMO nonlinear systems, containing backlash as an input nonlinearity. More specifically, a state observer is put forward to estimate non-measurable states. The control strategy was segmented into two modules: a type-2 fuzzy module that approximates the uncertainty in the model, and an H-infinity-based controller module that reduces the effects of external perturbations.

In [2], a stable robust adaptive interval type-2 fuzzy H-2/H-infinity, controller (RAIT2FH(2)H(infinity)C) for a class of uncertain nonlinear systems was proposed, which aims to address the above concerns through its hybrid robust/adaptive structure. More specifically, the H-2 energy and tracking function is optimized with respect to a H-infinity, perturbation attenuation constraint, while the interval fuzzy system manages the uncertainty in the approximation of the unknown system dynamics. In [13], an interval type-2 fuzzy logic controller is designed for under actuated truss-like robotic finger to accomplish the goal of stabilization in its equilibrium point. In addition, the behaviors of the controller are compared with respect to the type-1 fuzzy case to highlight the advantages of the type-2 approach.

In [10], a method for finding the optimal value of n in type-n fuzzy systems is presented. Also, in this work, the author proposed the utilization of bio-inspired optimization algorithms with fuzzy dynamic parameter adaptation for obtaining the optimal n value. The proposed approach may be utilizable in optimally designing type-3 (or higher type) fuzzy controllers in robotics, which could potentially (in the near future) achieve higher performance than type-2 fuzzy control. More in the long-term future, it is possible that type-4 fuzzy or even higher type-n could be applicable to real problems in diverse areas, such as control, pattern recognition, medical diagnosis, and time series prediction.

An optimal interval type-2 fuzzy logic control-based closed-loop drug administration to regulate the mean arterial blood pressure is presented in [42] which consisting of an interval type-2 fuzzy controller, which acts as pre-compensator to the traditional PID controller is presented, to regulate the mean arterial blood pressure of a patient by administering the drug sodium nitroprusside in a controlled manner. An effective nature-inspired optimization technique, which is called the cuckoo search algorithm, is utilized for finding the optimal method parameters.

In [17], the authors designing an interval type-2 fuzzy pre-compensated PID controller applied to two-DOF robotic manipulator with variable payload. Also, (Nodeh, Ghasemi, and Daniali 2019b) were proposed a method based on tuning of the higher-order sliding mode controller parameters by the interval type-2 fuzzy logic. In this case, the usual chattering effect of the traditional sliding mode is vanished by using higher-order sliding mode control and a saturation function that produces higher robustness when compared to the traditional method.

In [24], a novel application of artificial bee colony (ABC) algorithm to optimize the fractional-order operators and scaling factors of interval type-2 fractional-order fuzzy proportional integral derivative controller was proposed for redundant robotic system for trajectory track controller structure is also implemented for benchmark fractional-order plants and uncertain inverted pendulum system.

A robot with the leader–follower approach has a disadvantage in the case of failure in formation if the leader robot fails. To overcome such problem, in [47], the authors proposed the formation control using interval type-2 fuzzy logic controller (IT2FLC). In this case, to validate the appropriateness of the controller, several simulations were undertaken under different environmental situations with several different parameters.

In [3] were proposed a stable indirect adaptive robust mixed H-2/H-infinity control approach to an alpha-plane representation of a general type-2 fuzzy framework. The model is shown to be both efficient and very useful for performing a theoretical analysis. In [20], an improved nonstationary fuzzy system approach versus type-2 fuzzy system for the lifting motion control with human in the loop simulation was presented.

Finally, in [29], proposal of a novel application of biogeography-based optimization (BBO) to optimize a controller in order to achieve high performance estimation of states. The type-2 fuzzy approach is utilized to eliminate the chattering problem in the sliding mode control. The Lyapunov theorem is used to establish a proof of stability for the controller. The novel control approach is validated utilizing a computational simulation of a two-link robot arm problem with very good results.

The previous works can be viewed as a representative sample of the research that has been done in recent years in the area of fuzzy logic for robotics.

3 Review of Type-2 Fuzzy Logic in Robotics

In this section, we are offering a review of recent works utilizing type-2 fuzzy logic in robotics with the goal of improving the understanding of this area, as well as envisioning its possible evolution. In the literature review, we presented several articles in a general way using type-2 fuzzy for robotics. Based on the search that was done in Web of Science "Type-2 fuzzy logic systems in Robotic", we encountered 19 works mentioned above. In Figs. 3 and 4, we can observe as in recent years the use of type-2 fuzzy in robotics is increasing. According to data collected in WoS and using the software VOSviewer [15], the network is shown in different views in Figs. 5 and 6 obtaining only six connected authors, the type of analysis to create the map was made by co-authorship and the unit of analysis is by authors.

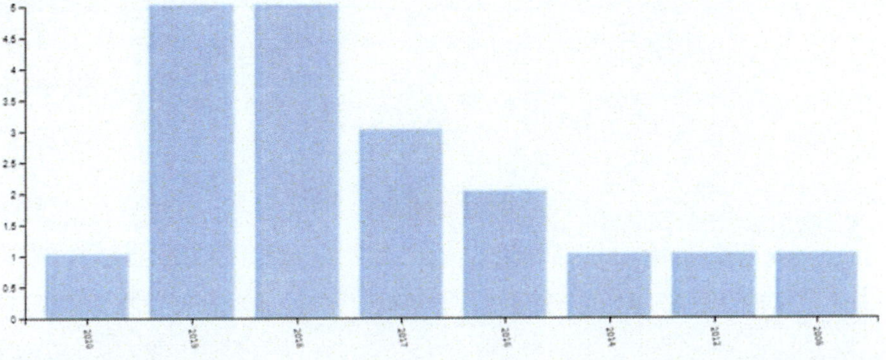

Fig. 3 Showing 19 records with number of papers by year for the TOPIC: (Type-2 fuzzy logic systems in robotic)

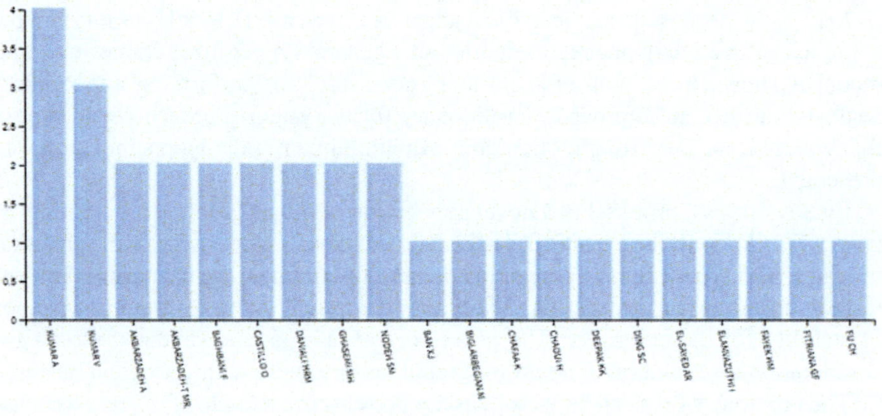

Fig. 4 Showing 19 records with the main authors and their papers for the TOPIC: (Type-2 fuzzy logic systems in robotic)

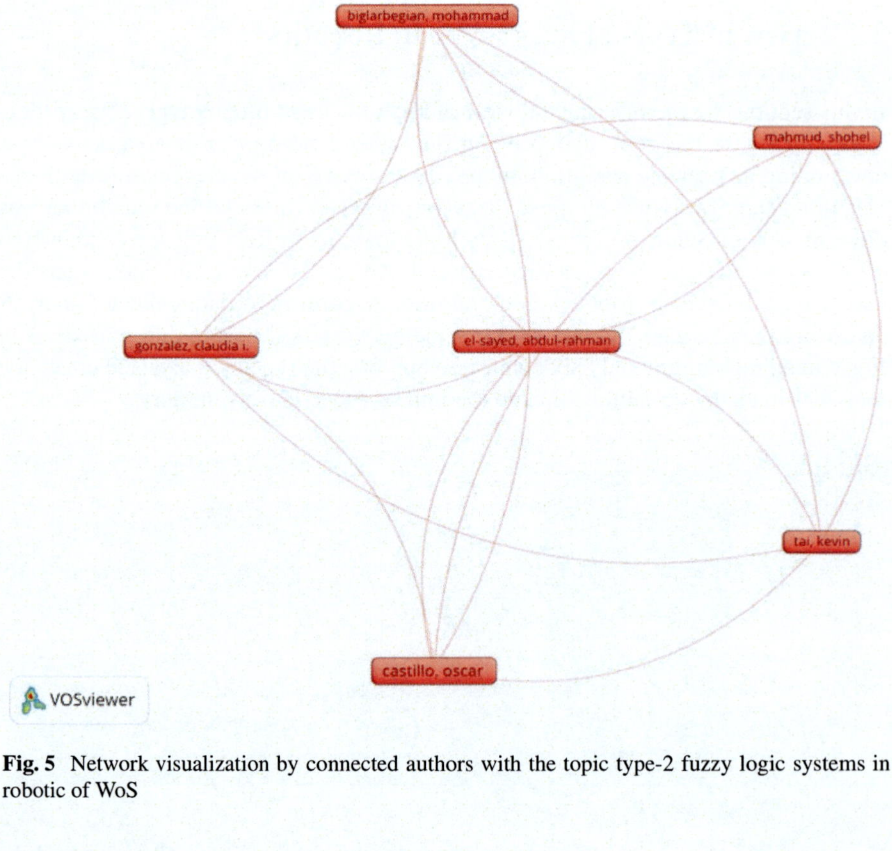

Fig. 5 Network visualization by connected authors with the topic type-2 fuzzy logic systems in robotic of WoS

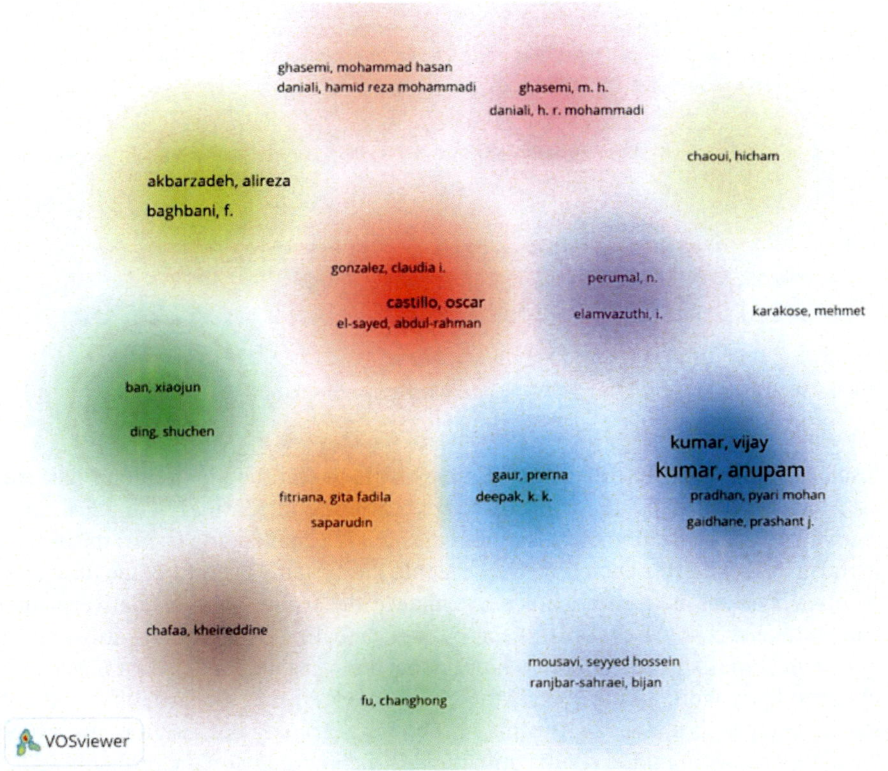

Fig. 6 Cluster density for all authors with the topic'type-2 fuzzy logic systems in robotic' of WoS

4 Review of Type-1 Fuzzy Logic in Robotics

In this section, we are offering a summary of the recent works using type-1 fuzzy logic in robotics with the goal of improving the understanding of this area. Based on the search that was done in Web of Science "Fuzzy logic systems in Robotics", we encountered 422 works. However, the first eight papers are cited in this review but the complete list can be consulted in WoS with the topic above mentioned ([7, 27, 21, 34, 41, 45, 43, 46]). In Fig. 7, we can observe as in the last years the use of type-1 fuzzy logic in robotic is relatively greater than type-2 fuzzy systems, but expect in the future that type-2 and type-3 will increase at a greater rate.

5 Conclusions

After finalizing the review, we can note that a plethora of articles utilizing fuzzy systems in robotics have been put forward with the goal of improving the navigation

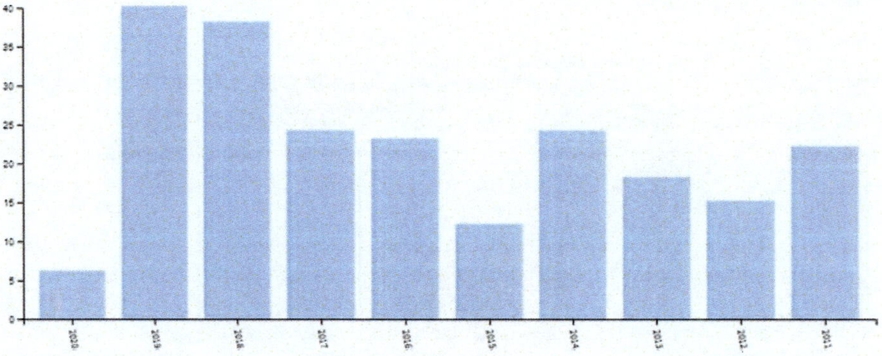

Fig. 7 Showing 422 records for TOPIC: (Fuzzy logic systems in robotics)

results of the robots. The relevance of fuzzy logic in real robotic problems has been established. Nowadays, the controllers are becoming increasingly complex, and a learning process for parameters is required, and it is a challenging task to the users to perform trial and error each time they execute the system. Therefore, the utilization of fuzzy systems is a good choice to enhance the results with a relatively small computational cost. Also, with this survey, we can realize the increasing utilization of type-1 and type-2 fuzzy systems, which are becoming more popular. With the results obtained from WoS, and the visual tool 'connected papers', we have observed how the tendency is to use fuzzy systems in robotics every day because the researchers have been achieving good results utilizing this methodology on the problems. Also, nowadays, we can find in the state-of-the-art hybrid methods to enhance the results. Also, as a future work, we can explore more combinations, for example, optimization methods using fuzzy logic for parameter adaptation could be included in the survey of hybrid methods with parameter adaptation using type-1 or type-2 fuzzy logic for comparison among the methods. Another advantage of elaborating this type of review paper is the encouragement offered to the authors to utilize techniques to improve the performance achieving the best possible results with these techniques. Finally, we can mention that a more recent related area that has been receiving increasing attention is the use of type-3 fuzzy logic, and at the moment, there are only a few papers on type-3 fuzzy control and robotics, but believe that this new area will gain more popularity in the near future. More in the long-term future, it would be possible to talk of type-4 fuzzy logic or even higher types of fuzzy applied to intelligent control and robotics.

References

1. Al-Rikabi, W.R.I.: Circumventing the fuzzy type reduction for autonomous vehicle controller (2017)
2. Baghbani, F., Akbarzadeh-T, M.-R., Akbarzadeh, A., Ghaemi, M.: Robust adaptive mixed H2/H∞ interval type-2 fuzzy control of nonlinear un- certain systems with minimal control effort. Eng. Appl. Artif. Intell. **49**(03), 88–102 (2016)
3. Baghbani, F., Akbarzadeh-T, M.R. and Akbarzadeh, A.: Indirectadaptive robust mixed H 2/H ∞ general type-2 fuzzy control of uncertain nonlinear systems. Appl. Soft Comput. **72**(08) (2018)
4. Bernal, E., Castillo, O., Soria, J., Valdez, F.: Optimization of fuzzy controller using galactic swarm optimization with type-2 fuzzy dynamic parameter adjustment. Axioms **8**(1), 26 (2019)
5. Bernal, E., Lagunes, M.L., Castillo, O., Soria, J. and Valdez, F.: Interval type-2 fuzzy logic for dynamic parameter adjustment in the imperialist competitive algorithm. In: 2019IEEE International Conference on Fuzzy Systems, {FUZZ-IEEE} 2019, New Orleans, LA, USA. IEEE, pp. 1–5 (2019)
6. Biglarbegian, M.: Systematic design of type-2 fuzzy logic systems for modeling and control with applications to modular and reconfigurable robots, Ph.D. Thesis, University of Waterloo, Canada (2010)
7. Blondin, M., Pardalos, P.: A holistic optimization approach for inverted cart-pendulum control tuning. Soft Comput. 24, 4343–4359 (2020)
8. Caraveo, C., Valdez, F., Castillo, O.: A new meta-Heuristics of optimization with dynamic adaptation of parameters using type-2 fuzzy logic for trajectory control of a mobile robot. Algorithms **10**(3), 85 (2017)
9. Castillo, O., Melin, P., Valdez, F., Soria, J., Ontiveros-Robles, E., Peraza, C., Ochoa, P.: Shadowed Type-2 fuzzy systems for dynamic parameter adaptation in harmony search and differential evolution algorithms. Algorithms **12**(1), 17 (2019)
10. Castillo, O.: Towards finding the optimal n in designing type-n fuzzy systems for particular classes of problems: a review. Appl. Comput. Math. 17(1):3–9 (2018)
11. Chaoui, H., Gueaieb, W.: Type-2 fuzzy logic control of a flexible-joint manipulator. J. Intell. Robot. Syst. JIRS **51**(02), 159–186 (2008)
12. Deepa, T.: Control of an Ambiguous real time system using interval type 2 fuzzy logic control (2017)
13. Ding, S., Huang, X., Ban, X., Lu, H. and Zhang, H.: Type-2 fuzzy logic control for underactuated truss-like robotic finger with comparison of a type-1 case1 (2017)
14. Du, X., Ying, H.: Derivation and analysis of the analytical structures of the interval type-2 Fuzzy-PI and PD controllers. IEEE Trans. Fuzzy Syst. **18**(4), 802–814 (2010)
15. Eck, N.J.V. and Waltman, L.: Visualizing bibliometric networks. In: Ding, Y., Rousseau, R., Wolfram, D. (eds.) Measuring Scholarly Impact: Methods and Practice, pp. 285–320. Cham: Springer International Publishing (2014)
16. Fayek, H.M., Elamvazuthi, I., Perumal, N., Venkatesh, B.: A controller based on optimal type-2 fuzzy logic: systematic design, optimization and real-time implementation. ISA Trans. **53**(5), 1583–1591 (2014)
17. Gaidhane, P.J., Nigam, M.J., Kumar, A., Pradhan, P.M.: Design of interval type-2 fuzzy precompensated PID controller applied to two-DOF robotic manipulator with variable payload. ISA Trans. **89**, 169–185 (2019)
18. Galluzzo, M., Cosenza, B.: Adaptive type-2 fuzzy logic control of a bioreactor. Chem. Eng. Sci. **65**(14), 4208–4221 (2010)
19. Hagras, H.: A type-2 fuzzy logic controller for autonomous mobile robots. In: 2004 IEEE International Conference on Fuzzy Systems (IEEE Cat. No.04CH37542), vol 2 (2004)
20. Karakose, M.: An Improved nonstationary fuzzy system approach versus type-2 fuzzy system for the lifting motion control with human-in-the-loop simulation. Int. J. Comput. Intell. Syst. **11**(01), 183 (2018)

21. Khalifeh, A., Rajendiran, K., Darabkh, K.A., Khasawneh, A.M., AlMomani, O. and Zinonos, Z.: On the potential of fuzzy logic for solving the challenges of cooperative multi-robotic wireless sensor networks (2019)
22. Kumar, A., Kumar, V.: A novel interval type-2 fractional order fuzzy PID controller: design, performance evaluation, and its optimal time domain tuning. ISA Trans. **68**, 251–275 (2017)
23. Kumar, A., Kumar, V.: Evolving an interval type-2 fuzzy PID controller for the redundant robotic manipulator. Expert Syst. Appl. **73**, 161–177 (2017)
24. Kumar, A., Kumar, V.: Design of Interval Type-2 Fractional-Order Fuzzy Logic Controller for Redundant Robot with Artificial Bee Colony. Arab J Sci Eng **44**, 1883–1902 (2019)
25. Lagunes, M.L., Castillo, O., Valdez, F., Soria, J.: Multi-Metaheuristic competitive model for optimization of fuzzy controllers. Algorithms **12**(5), 90 (2019)
26. Likaj, R., Bajrami, X., Shala, A., Pajaziti, A.: Path finding for a mobile robot using fuzzy and genetic algorithms. Int. J. Mech. Eng. Technol. **8**(01), 659–669 (2017)
27. Ling, S., Wang, H., Liu, P.X.: Adaptive fuzzy tracking control of flexible- joint robots based on command filtering. IEEE Trans. Industr. Electron **67**(5), 4046–4055 (2020)
28. Mahapatra S.: Analysis and control of mobile robots in various environmental conditions (2012)
29. Medjghou, A., Ghanai, M., Chafaa, K.: BBO optimization of an EKF for interval type-2 fuzzy sliding mode control. Int. J. Comput. Intell. Syst. **11**(03), 770 (2018)
30. Mendel, J.M.: Advances in type-2 fuzzy sets and systems. Inf. Sci. **177**(1), 84–110 (2007)
31. Mendel, J.M.: Type-2 fuzzy sets as well as computing with words. IEEE Comput. Intell. Mag. **14**(1), 82–95 (2019)
32. Mousavi, S., Ranjbar-Sahraei, B., Noroozi, N.: Output feedback controller for hysteretic time-delayed MIMO nonlinear systems. Nonlinear Dynam **68**(04) (2012)
33. Musikasuwan, S.:Novel fuzzy techniques for modelling human decision making (2013)
34. Muthugala, M.V.J., Vengadesh, A., Wu, X., Elara, M.R., Iwase, M., Sun, L. and Hao, J.: Expressing attention requirementof a floor cleaning robot through interactive lights. Autom. Constr. **110**:103015
35. Nguyen, N.T., Núñez, M., Trawinski, B.: Collective intelligent information and database systems. J. Intell. Fuzzy Syst. **32**(12), 1–4 (2016)
36. Nodeh, S., Ghasemi, M., Daniali, H.R.: Hybrid robust controller based on interval type 2 fuzzy neural network and higher order sliding mode for robotic manipulators. Nexo Revista Científica **32**(12), 106–125 (2019)
37. Nodeh, S.M., Ghasemi, M.H., Mohammadi Daniali, H.R.: Robust tuned controller based on interval type 2 fuzzy logic for robotic manipulators exposed to perturbations and parametric uncertainties. J. Control Autom Electr. Syst. **30**(3), 323–336 (2019)
38. Ortiz, Oscar Rigoberto Carvajal, Oscar Castillo, andSoria, J.: Optimization of membership function parameters for fuzzy controllers of an autonomous mobile robot using the flower pollination algorithm. J. Autom. Mobile Robot. Intell. Syst.**12**(03) (2018)
39. Sakinah, S., Syed, W. A.: A comparison of type-1 and type-2 fuzzy logic controllers in robotics: a review (2015)
40. Sarabakha, A., Fu, C., Kayacan, E. and Kumbasar, T.: Intuit before tuning: type-1 and type-2 fuzzy logic controllers. Appl. Soft Comput. **81**:105495
41. Seçkin, A.Ç., Karpuz, C. and Özek, A.: Comparisonof cooperative behaviors in multiple robots and fuzzy logic approach (2019)
42. Sharma, R., Deepak, K.K., Gaur, P., Joshi, D.: An optimal interval type-2 fuzzy logic control based closed-loop drug administration to regulate the mean arterial blood pressure. Comput. Methods Programs Biomed. **185**(10), 105167 (2019)
43. Sun, D., Kiselev, A., Liao, Q., Stoyanov, T., Loutfi, A.: A new mixed-reality-based teleoperation system for telepresence and maneuverability enhancement. IEEE Trans. Hum. Mach. Syst. **50**(1), 55–67 (2020)
44. Tai, K., El-Sayed, A.R., Biglarbegian, M., Gonzalez, C.I., Castillo, O. and Mahmud, S.: Review of recent type-2 fuzzycontroller applications. Algorithms 9(2): 39 (2016)
45. Toffano, Z., Dubois, F.: Quantum eigenlogic observables applied to the study of fuzzy behaviour of Braitenberg vehicle quantum robots. Kybernetes **48**(10), 2307–2324 (2019)

46. Tucan, P., Gherman, B., Major, K., Vaida, C., Major, Z., Plitea, N., Carbone, G., Pisla, D.: Fuzzy logic-based risk assessment of a parallel robot for elbow and wrist rehabilitation. Int. J. Environ. Res. Public Health **17**(01), 654 (2020)
47. Tutuko, B., Nurmaini, S., Saparudin, S., Fadila, G.: enhancement of non-holonomic leader-follower formation using interval type-2 fuzzy logic controller. Int. J. Online Eng. (iJOE) **14**(09), 124 (2018)
48. Ulu, C., Güzelkaya, M., Eksin, I.: A dynamic defuzzification method for interval type-2 fuzzy logic controllers. IEEE Int. Conf. Mechatron. 318–323
49. Valdez, F., Castillo, O., Caraveo, C., Peraza, C.: Comparative study of the conventional mathematical and fuzzy logic controllers for velocity regulation. Axioms **8**(2), 53 (2019)
50. Zadeh, L.A.: The concept of a linguistic variable and its application to approximate reasoning—I. Inf. Sci. **8**(3), 199–249 (1975)
51. Zhang, W., Xing, Y., Qiu, D.: On characterizations of directional derivatives and subdifferentials of fuzzy functions. Symmetry 9(09) (2017)

Optimization of Palm Oil Mill Effluent (POME) Solubilization Using Linguistic Fuzzy Logic and Machine Learning Techniques

Zuzana Jankova, Petr Dostal, Dipak Kumar Jana, Samyabrata Bhattacharjee, Barnali Bej, Priyanka Dey, and Sudipta Roy

Abstract The continuous hike in the price of the edible vegetable oil has directly impacted upon the price growth of the palm oil. In this palm oil production, Indonesia and Malaysia are in top leading position at present. It is found that an increase in palm oil mill setup is radically increasing the produce of effluent discharge which is a severe threat to our environment. Hence to maintain the balance of our already affected ecosystem, the proper treatment of this residual product is becoming the dire need of the hour. The conventional method used for solubilization of palm oil mill effluent (POME) is thermal alkaline pre-treatment. In this paper, the linguistic fuzzy logic (LFL) and machine learning (ML) techniques have been used to analyze the data, and type-2 fuzzy logic controller (T2FLC) has been used to optimize the solubilization of POME. The effect of reaction time, NaOH concentration, and temperature on the solubilization has been evaluated. From the investigation of the surface plot, which developed in T2FLC environment, it has been observed that NaOH concentration has a significant effect on the solubilization of POME. The prediction efficiency of T2FLC then has been compared with T1FLC and RSM. By evaluating some statistical analyses, the sensitivity and validity of the proposed model have finally been measured.

Keywords Palm oil mill effluent · Type-2 fuzzy logic · Solubilization · Thermo-alkaline pre-treatment · Statistical analysis

Z. Jankova · P. Dostal
Faculty of Business and Management, Brno University of Technology, Brno, Czech Republic
e-mail: Zuzana.Jankova@vutbr.cz

P. Dostal
e-mail: dostalp@vut.cz

D. K. Jana · P. Dey · S. Roy
School of Applied Science and Humanities, Haldia Institute of Technology, Haldia, West Bengal, India
e-mail: dipakjana@gmail.com

S. Bhattacharjee · B. Bej (✉)
Department of Chemical Engineering, Haldia Institute of Technology, Haldia 721657, India
e-mail: barnalibej@gmail.com

O. Castillo et al. (eds.), *Applied Mathematics and Computational Intelligence*,
Springer Proceedings in Mathematics & Statistics 413,
https://doi.org/10.1007/978-981-19-8194-4_18

1 Introduction

Palm oil is the main outcome from the agricultural zone, mainly obtained in the tropical regions. Indonesia and Malaysia are the leading hub in producing palm oil. Palm oil is obtained from the fruit of oil palm tree scientifically known as Elaeis guineensis. Palm oil is extensively used as raw material in the various food and non-food industries like cosmetics and pharmaceutical industry. An increase in population continuously increases the demand for food production from the agricultural sector which will make it the most water-polluting sector [1]. The gradual increase in the temperature range and frequent rainfall has increased the requirement of the water by the agriculture zones to meet the demand of the growing population [2]. The obstruction that is faced by palm oil production firm is the palm oil mill effluent, which is the waste product. According to some statistics, each tonne of palm oil production generates nearly 2.5–3.0 cubic meter of POME [3]. POME is the thick brownish color colloidal suspension whose characteristics are given in Table 1. The characteristics of the effluent vary which depend on the operational process used in mill for palm oil production. It has high value of COD and BOD which are present nearly 50,000 ppm and 25,000 ppm, respectively [4, 5]. The use of POME as a feedstock has picked up the enthusiasm of scientists to control squander creation in agriculture division obtained from the palm oil mill industry. Complex techniques have been widely evolved to treat and to use the waste product which can be categorized into four primary sections: biological, thermochemical, physiochemical, and the mix of the techniques thereof. Some of the techniques are dealt only to solve the waste-related problems, while part of them means to recoup vitality from the treatment. Each techniques has a few favorable circumstances and burdens, which is important before implementing it. Because of the degradable organic constituent, wastewaters have the potential to be utilised, where a net positive vitality addition could be accomplished with legitimate methodology [6].

For palm oil mill effluent, anaerobic digestion (AD) is commonly used by industries since it is a most cost-effective and ecofriendly process [16]. It is simply defined as the series of process in which the organic component breaks down into bio-energy in the absence of oxygen. The process of generation of bio-energy from anaerobic digestion can be divided into four sections: hydrolysis, acidogenesis, acetogenesis, and methanogenesis [17]. These techniques have several advantages as follows [18]:

- Anaerobic digestion differs from aerobic digestion in terms of availability of oxygen. Thus, the anaerobic digestion reduces the energy cost for the extensive supply of oxygen.
- Volatile solids content reduces through this operation. As per some literature, the addition of nitrite enhances the reduction of volatile suspended solid and increases the sludge treatment efficiency [19].
- By this operation, energy is recovered by the production of methane.
- It kills large percentage of pathogens present in sludge.

Among different pre-treatment methodology, the most well-known and broadly executing process is thermo-alkaline pre-treatment technique. In this paper, this

Table 1 Literature comparison of anaerobic digestion of different waste product

Waste	Pretreatment method	Experimental result	Reference
Microbial biomass	pH=8.5, Temperature $= 35°C$	Solubilization is 50.6%	[7]
Waste-activated sludge	7g/l NaOH concentration	Solubilization is 43.5%	[8]
Maize cob	microwave irradiation	Biogas yield increased by 46%	[9]
Wheat straw	Hydrogen peroxide	Methane yield increase by 50%	[10]
Waste-activated sludge	pH = 12 at ambient temperature	Solubilization is 30.7%	[11]
Brewery wastewater	Mesophilic at temperature 35°C	Methane yield efficiency is 0.52 to 0.53	[12]
POME	Ozonation method pH = 7.2	Methane yield is 64.1410 LCH_4/kg COD	[13]
POME	Ultrasonic method pH = 7.2 Temperature=32 °C to 37°C	Methane yield is 44 LCH_4/kg COD	[14]
POME	Temperature = 32.5 °C C time = 41.23 hr 8.83g/l of NaOH concentration	Solubilization	[15]
POME	Temperature = 35 °C Incubation time = 48 hr 8 g/l of NaOH concentration	Solubilization is 82.7%	This study

pre-treatment methodology is selected for solubilization and hydrolysis of macromolecule into simple monomers. The surface area and the rate of hydrolysis process increase due to the reduction of particle size which reduce the reaction time of the hydrolysis operation [20]. As per some previous study, sodium hydroxide (NaOH), calcium hydroxide (Ca(OH)2), and potassium hydroxide (KOH) are widely used as alkali in this pre-treatment process [21]. As per some study, the efficiency of sludge solution varies with the selection of alkali in a order of NaOH > KOH > $Mg(OH)_2$ and $Ca(OH)_2$ [8].

To increase the COD solubilization % of the effluent by thermo-alkaline pre-treatment methodology is optimized with respect to incubation time, NaOH concentration, and temperature by various soft computing techniques. It is frequently called as computational insight, covering a scope of strategies in the field of computer science, machine learning and AI. It is the collection of techniques that are dealt with the uncertain problem and provide low-cost optimum solution [22]. Soft computing techniques comprise artificial neural networks (ANN), fuzzy logic (FL) [23], adaptive neuro-fuzzy inference systems (ANFIS) [24], genetic algorithms (GAs) , data mining (DM), etc. The soft computing techniques used for the optimization the solubilization % of effluent obtain from palm oil mill by techniques like response surface methodology (RSM) [15] and fuzzy logic (FL) which is developed in this paper.

The type-1 fuzzy logic (T1FLC) is first proposed by Zadeh [25]. The most famous sort 1 fuzzy inference (T1FI) models are proposed by Mamdani and Assilian ([26]) and Sugeno (Takagi and Sugeno [27]). A sort 2 fuzzy set (T2FS) has grades of enrollment that are type-1 fuzzy [25, 28–34], so it very well may be known as a $fuzzyfuzzyset$, and in this way, IT2-FLS has the additional kind decrease measure. Therefore, the type-2 fuzzy logic controller (T2FLC) is widely used to handle uncertain and nonlinear environment more efficiently by making use of fuzzy sets [35–40]. Brief mathematical detailing is provided in Sect. 3

We have developed the following in this optimizing work:

- Optimization of COD solubilization % by T2FLC model is done.
- Influence of the independent parameter, time, temperature, and NaOH concentration is investigated.
- Legitimacy of the proposed model is finished by statistical analysis.
- The prediction efficiency of the proposed model, T2FLC, is compared with other soft computing techniques such as RSM and T1FLC.

The graphical abstract of the the proposed model is depicted in Fig. 1.

2 Type-2 Fuzzy Sets

A type-2 fuzzy set (T2FS) communicates by the non-deterministic truth degree having imprecision and weakness for a part that has a spot with a set [41]. A type-2 fuzzy set (T2FS) denoted by $\tilde{\tilde{A}}$ [42] is portrayed by a type-2 membership function (T2MF) $\mu_{\tilde{\tilde{A}}}(s,t)$ where $s \in X, \forall t \in J_s^t \subseteq [0,1]$ and $0 \leq \mu_{\tilde{\tilde{A}}}(s,t) \leq 1$ are defined in equation (1)

$$\tilde{\tilde{A}} = \{(s, t, \mu_{\tilde{\tilde{A}}}(s,t)) | s \in X, \forall t \in J_s^t \subseteq [0,1]\} \tag{1}$$

If $\tilde{\tilde{A}}$ is fuzzy type-2 (FT2) continuous variable, it is denoted in Eq. (2)

$$\tilde{\tilde{A}} = \left\{ \int_{s \in X} \left[\int_{t \in J_s^t} f_s(t)/t \right] /s \right\} \tag{2}$$

where $\int\int$ denotes as the union of s and t. If A is FT2 discrete, then it is denoted by equation (3)

$$\tilde{\tilde{A}} = \left\{ \sum_{s \in X} \mu_{\tilde{\tilde{A}}}(s)/s \right\} = \left\{ \sum_{i=1}^{N} \left[\sum_{k=1}^{M_i} f_{s_i}(t_k)/t_{ik} \right] /s_i \right\} \tag{3}$$

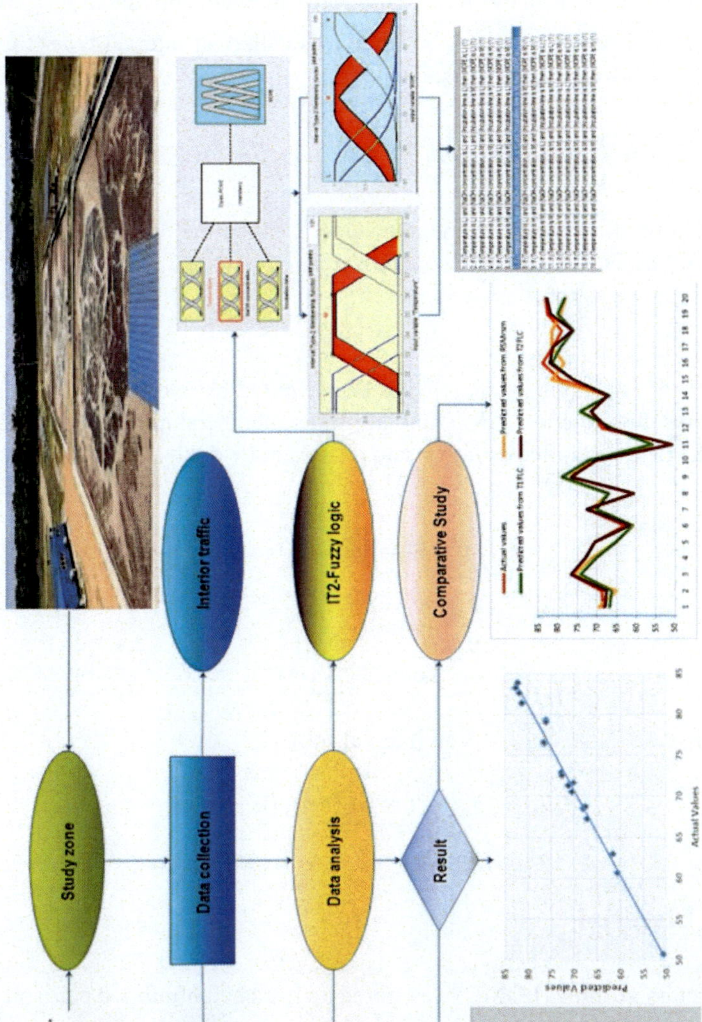

Fig. 1 Graphical abstract of the proposed model

where $\sum\sum$ denotes the union of s and t. If $f_s(t) = 1, \forall t \in [\underline{J}_s^t, \overline{J}_s^t] \subseteq [0, 1]$, the T2MF $\mu_{\tilde{\tilde{A}}}(s, t)$ is expressed by one type-1 inferior membership function, $\underline{J}_s^t = \mu_A(s)$ and one type-1 superior, $\overline{J}_s^t = \mu_A(s)$, then it is called an interval type-2 fuzzy set (IT2FS) defined by equations (4) and (5).

$$\tilde{\tilde{A}} = \left\{(s, t, 1) | \forall s \in X, \forall t \in [\underline{\mu}_A(s), \overline{\mu}_A(s)] \subseteq [0, 1]\right\} \tag{4}$$

The union of all the primary memberships is called the footprint of uncertainty (FOU) of $\tilde{A}$. The FOU$\tilde{A}$ can be characterized as

$$\text{FOU}(\tilde{A}) = \cup_{\forall s \in X} J_s = (s, t) : t \in J_s \subseteq [0, 1] \tag{5}$$

The FOU of type-2 fuzzy set ($\tilde{A}$) has been limited by two type-1 MFs called as lower membership function (LMF) and the upper membership function (UMF). The UMF and LMF are signified as $\overline{\mu}_{\tilde{A}}$ (s) and $\underline{\mu}_{\tilde{A}}$ (s) , individually, and are characterized as follows:

$$\overline{\mu}_{\tilde{A}}(s) = \overline{\text{FOU}(\tilde{A})} \tag{6}$$

and,

$$\underline{\mu}_{\tilde{A}}(s) = \underline{\text{FOU}(\tilde{A})} \tag{7}$$

where $\forall_s \in X$. Note that J_s is an interval set, i.e.,

$$J_s = (s, t) : t \in [\underline{\mu_{\tilde{A}}}(s)], [\overline{\mu}_{\tilde{A}}(s)] \tag{8}$$

3 Raw Material

Palm oil factory gushing (POME) was stored up from neighborhood palm oil factory. The collected effluent was stored in very low temperature, at the range of 4–6°C. The reason for keeping it at a low temperature is to prevent microorganism deterioration. At the beginning of experiment, collected sample of POME was brought to room temperature. NaOH used for the experiment is of analytical grade, brought from MERCK.

3.1 Physicochemical Analyses and Experimental Method

Collected samples of POME are analyzed before anaerobic digestion, the total solid (TS), total suspended solid (TSS), volatile solids (VS), and volatile suspended solids (VSS) measured in accordance with standard methods used for the examination of water [43].

The pH of the sample POME is marked with a pH meter. The total chemical oxygen demand (TCOD) and soluble chemical oxygen demand (SCOD) were analyzed using the closed reflux colorimetric method. Characterization of the samples collected from mill is given in Table 1. The ratio of SCOD and TCOD is always used for evolution in the extent of hydrolysis reaction [44]. The COD solubilization % is calculated by Eq. (9) [8].

A series of tests is performed in this batch experiment under different criteria of the independent parameter, time, temperature, and NaOH concentration. In the experimental process, every flask contains 100 ml of the collected sample of POME, and a different quantity of NaOH added. The concentration range of NaOH used is given in Table 4. For achieving the anaerobic condition, every flask was purged with nitrogen gas and sealed with parafilm to make the system gastight. Finally, the series of samples is incubated for different time and temperature. The time and temperature range in operation is given in Table 4.

$$\text{COD Solubilization\%} = \frac{\text{Soluble Chemical oxygen demand after treatment}}{\text{Total Chemical Oxygen Demand after treatment}} \times 100 \qquad (9)$$

3.2 Experimental Design

Here, we have developed a triple-input and single-output type-2 fuzzy logic system for optimizing the solubilization of POME. The input parameters temperature, reaction time, and NaOH concentration are considered. The optimality of output from the pre-treatment of POME by the thermo-alkaline method depends on the extent of COD solubilization. Therefore, we consider this as an output parameter of the developed model. Range of all the independent parameter is considered for experimentation as well for optimization given in Table 2.

The structure of the Mamdani fuzzy inference system is depicted in Fig. 2. The input parameters, temperature, reaction time, NaOH concentration, and outer parameter used for the development of the model are divided into three linguistic terms, low (L), middle (M), and high (H), which is quickly processed by the type-2 fuzzy set theory lucidly. Interval type-2 fuzzy (IT2F) semantic factors are delays of mathematical factors as in they can address the state of a trait at a given interval by taking IT2F sets as their qualities. Membership function of input and output parameters is depicted in Figs. 3 and 4.

Table 2 Parametric analysis of raw POME sample

Parameter use for analyzing raw POME	Units	Ranges
Ph	–	4.1–5.0
Total solids-	mg/lit	34,110–60,800
Volatile solids	mg/lit	28,000–54,000
Total suspended solids	mg/lit	16,000–32,550
Volatile suspended solid	mg/lit	15,180–30,700
Total chemical oxygen demand	mg/lit	54,100–95,500
Soluble chemical oxygen demand	mg/lit	22,000–32,500

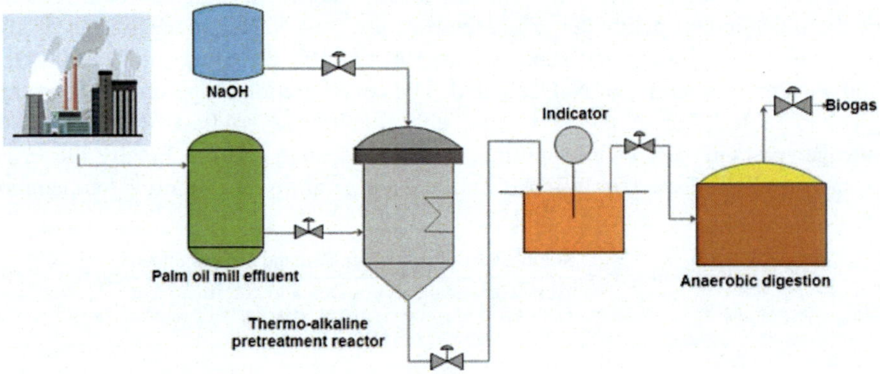

Fig. 2 Pre-treatment process diagram of the POME

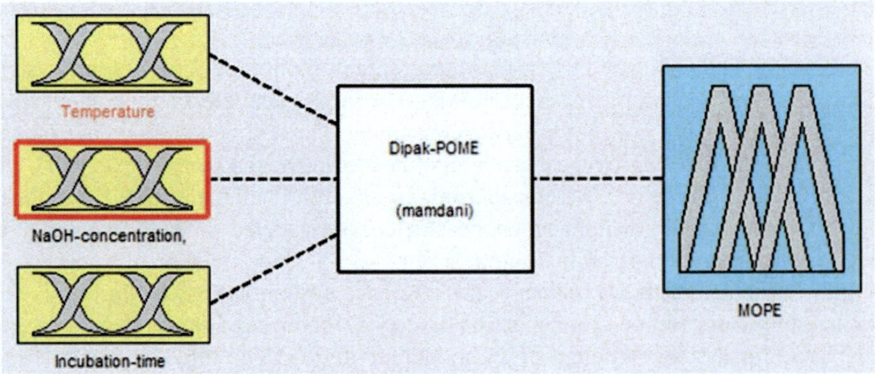

Fig. 3 Block diagram of the proposed model

Table 3 Ranges of parameter

Input parameter	Units	Labels				
		−1.682	−1	0	1	1.682
Temperature	C	26.6	30	35	40	43.4
NaOH concentration	g/L	1.27	4	8	12	14.73
Incubation time	h	7.64	24	48	72	88.36

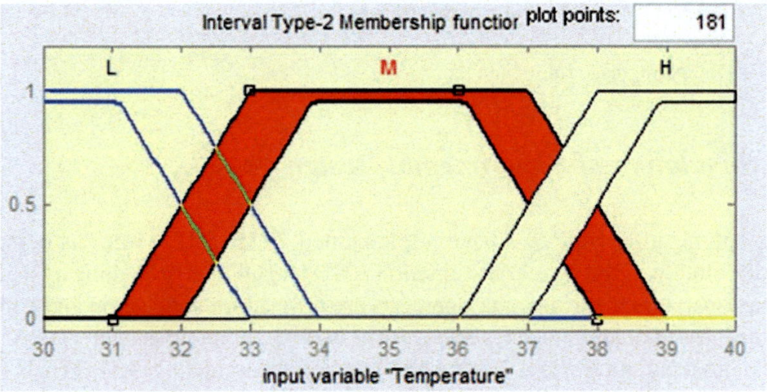

Fig. 4 Membership function of the input variable temperature

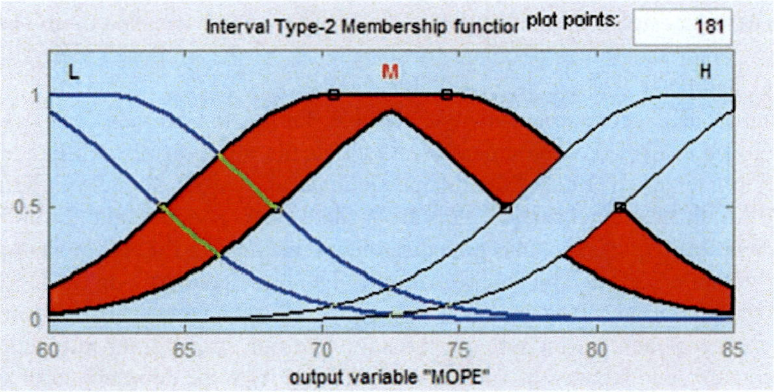

Fig. 5 Membership function of the input variable and COD solubilization%

1. If (Temperature is L) and (NaOH-concentration, is L) and (Incubation-time is L) then (MOPE is L) (1)
2. If (Temperature is L) and (NaOH-concentration, is L) and (Incubation-time is M) then (MOPE is L) (1)
3. If (Temperature is L) and (NaOH-concentration, is M) and (Incubation-time is L) then (MOPE is M) (1)
4. If (Temperature is L) and (NaOH-concentration, is M) and (Incubation-time is H) then (MOPE is H) (1)
5. If (Temperature is H) and (NaOH-concentration, is L) and (Incubation-time is L) then (MOPE is H) (1)
6. If (Temperature is H) and (NaOH-concentration, is H) and (Incubation-time is L) then (MOPE is M) (1)
7. If (Temperature is H) and (NaOH-concentration, is H) and (Incubation-time is M) then (MOPE is L) (1)
8. If (Temperature is L) and (NaOH-concentration, is M) and (Incubation-time is M) then (MOPE is M) (1)
9. If (Temperature is H) and (NaOH-concentration, is M) and (Incubation-time is M) then (MOPE is M) (1)
10. If (Temperature is M) and (NaOH-concentration, is L) and (Incubation-time is M) then (MOPE is L) (1)
11. If (Temperature is M) and (NaOH-concentration, is H) and (Incubation-time is M) then (MOPE is L) (1)
12. If (Temperature is M) and (NaOH-concentration, is M) and (Incubation-time is L) then (MOPE is M) (1)
13. If (Temperature is M) and (NaOH-concentration, is M) and (Incubation-time is L) then (MOPE is L) (1)
14. If (Temperature is M) and (NaOH-concentration, is M) and (Incubation-time is H) then (MOPE is L) (1)
15. If (Temperature is M) and (NaOH-concentration, is M) and (Incubation-time is M) then (MOPE is M) (1)
16. If (Temperature is M) and (NaOH-concentration, is M) and (Incubation-time is M) then (MOPE is H) (1)

Fig. 6 IF-THEN linguistic rule

3.3 Simulation of Experimental Design

In this optimization work, we have implemented 20 IF-THEN rules as depicted in Fig. 5. Evolution of the logical conjunction AND and OR has been done by using min and max operators. Min and max operators are utilized for suggestion and collection technique, individually, whose general form is If (first input parameter) is A1 AND (second input parameter) is $A_2, \cdots$ AND (mth input parameter) is A_m THEN (output parameter) is B, where $A_1, \cdots, A_m$ are the linguistic values of the respective input variable and B linguistic values of output variables. The last step is the defuzzifications step which is the interaction to changes over the fluffy yield of the induction motor to fresh esteem utilizing participation capacities are finished by the centroid strategy. Predicted values obtained by stimulation of the proposed model,T2FLC, T1FLC, and RSM [15] are given in Table 4 and plotted in Fig. 10.

From the data set of output and input variable, Table 4, three dimension surface plot is developed in T2FLC environment to investigate the variation of solubilization % of POME to interactive effect of the independent variable which is in the experimental range . All the generated surface plots are depicted in Figs. 6, 7, and 8. From Figs. 6, and 7, it is observed that NaOH concentration is considered being the most important factor among the three. The solubilization % increases gradually on the increase in alkali concentration in the process. According to some writing, expansion in the convergence of antacid increments the solubilization in different interaction like saponification and balance of various acids framed from the debasement of specific materials[8]. According to Fig. 8, the effect of incubation time on solubilization % is more dominate on temperature. At lower temperatures, more incubation time is required to attain higher % of solubilization Fig. 9.

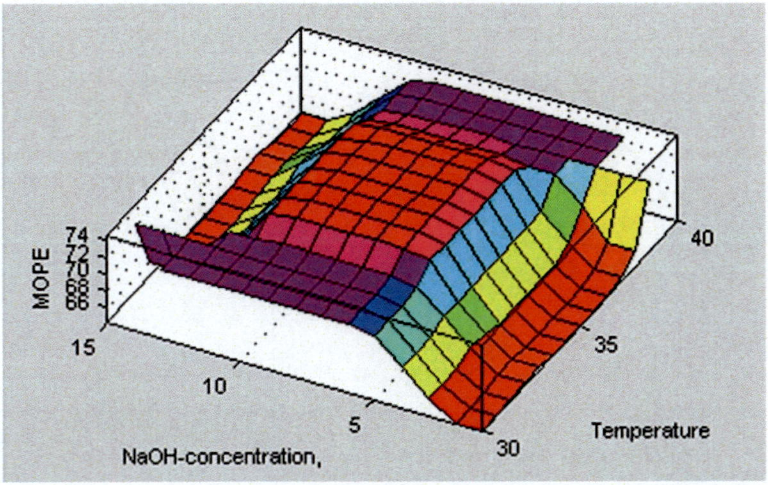

Fig. 7 Surface contour plot of COD solubilization% on NaOH concentration and temperature

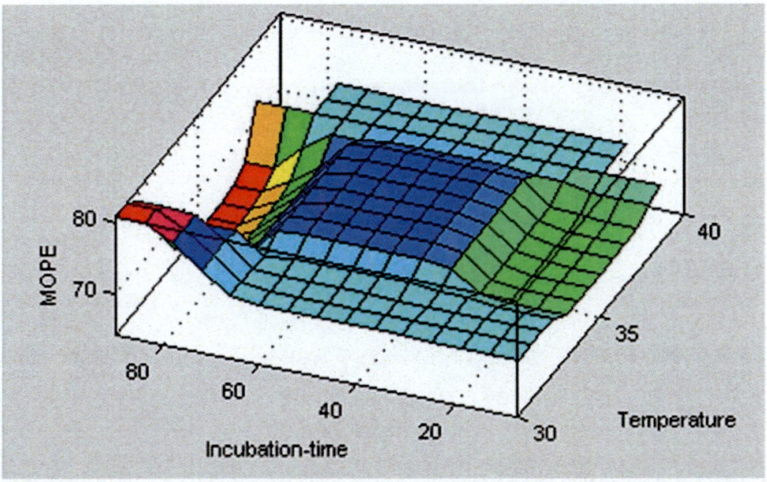

Fig. 8 Scatter plot between predicted values from T2FLC model and actual values of COD solubilization %

4 Results and Discussion

For every input, the proposed model is stimulated to obtain output results. Experimentally, getting the result is termed as actual value, and data obtained by stimulating the independent data set in the T2FLC model is termed as predicted values. The expectation capacity of the created model is determined by utilizing the test information in the prepared information and looking at the real qualities and anticipated qualities. In

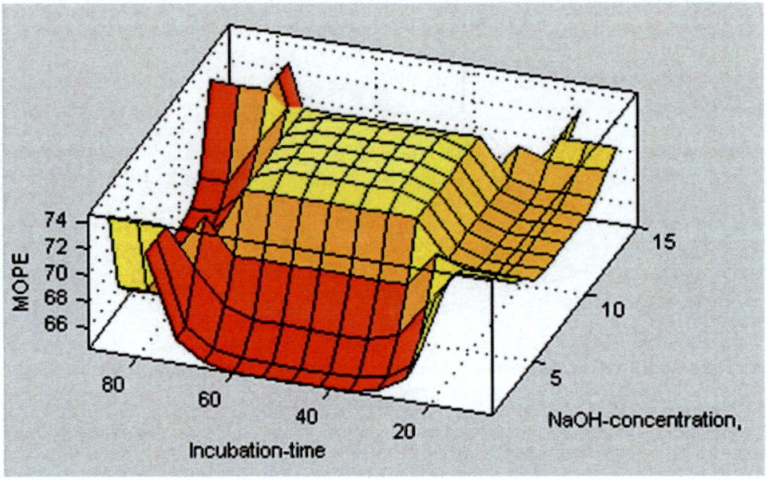

Fig. 9 Surface contour plot of COD solubilization % on incubation time and NaOH concentration

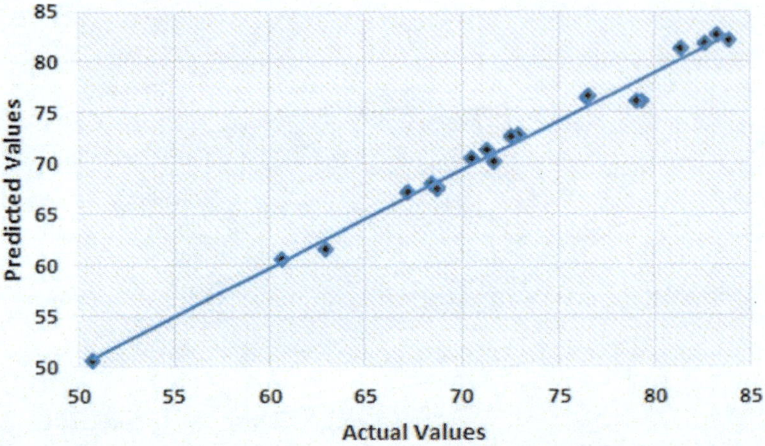

Fig. 10 Surface contour plot of COD solubilization % on temperature and incubation time

actual various predicted data set for COD solubilization depicted in Fig. 8 and Fig. 9. From the figure, it is observed that the predicted data set is obtained from T2FLC distributed near the straight line. For the measurement of the prediction capability, the statistical parameter is used. The statistical parameters are the root mean square error (RMSE), the determination coefficient (R^2), mean absolute percentage error (MAPE), and mean absolute error(MAE). All the statistical parameters are calculated by using Eq. (10), Eq. (11), Eq. (12), and Eq. (13), respectively, where n is the quantity of information designs in the informational index, y_{pred_i} demonstrates the anticipated worth, i is specific information point, and y_{act_i} is the real worth (which is depicted in Fig. 11).

Table 4 Statistical data analysis for T2FLC and T1FLC and RSM

FIS	RMSE	R^2	MAPE	MAE
T2FLC	0.104	0.991	4.703	0.072
T1FLC	0.056	0.953	3.965	0.039
RSM	0.051	0.923	3.121	0.032

$$\text{RMSE} = \sqrt{\frac{1}{n}\sum_{i=1}^{n}(y_{\text{pred}_i} - y_{\text{act}_i})^2} \tag{10}$$

$$R^2 = 1 - \frac{\sum_{i=1}^{n}(y_{\text{pred}_i} - y_{act_i})^2}{\sum_{i=1}^{n} y_{\text{act}_i}^2} \tag{11}$$

$$\text{MAPE} = \frac{1}{n}\sum_{i=1}^{n}\frac{|(y_{\text{pred}_i} - y_{\text{act}_i})|}{y_{\text{pred}_i}} \times 100\% \tag{12}$$

$$\text{MAE} = \frac{1}{n}\sum_{i=1}^{n}|y_{\text{pred}_i} - y_{\text{act}_i}| \tag{13}$$

The superiority of the proposed model can be predicted by analyzing the outcome of the statistical parameter. The closer the value R^2 to 1 and the smaller the value of RMSE, MAPE, and MAE, the greater the accuracy of the proposed model. The prediction capability of all the intelligent computing methods like RSM, T1FLC, and T2FLC is compared in terms of this statistical parameter, R^2, RSME, MAPE, and MAE. The outcome of all the parameters of all intelligent computing methods is given in Table 3 and Fig. 11.

5 Potential Application and Challenges

As per some literature, the solubilization process is considered to be beneficial for improving the anaerobic digestion rate and production of methane [8, 45]. Higher the generation of methane, more is the production of renewable energy. The biggest obstruction faced by the palm oil mill all over the world is a continuous threat to the ecosystem. The extensive treatment system is the traditional method widely used for pre-treatment POME. There are a couple of regions that value abuse in

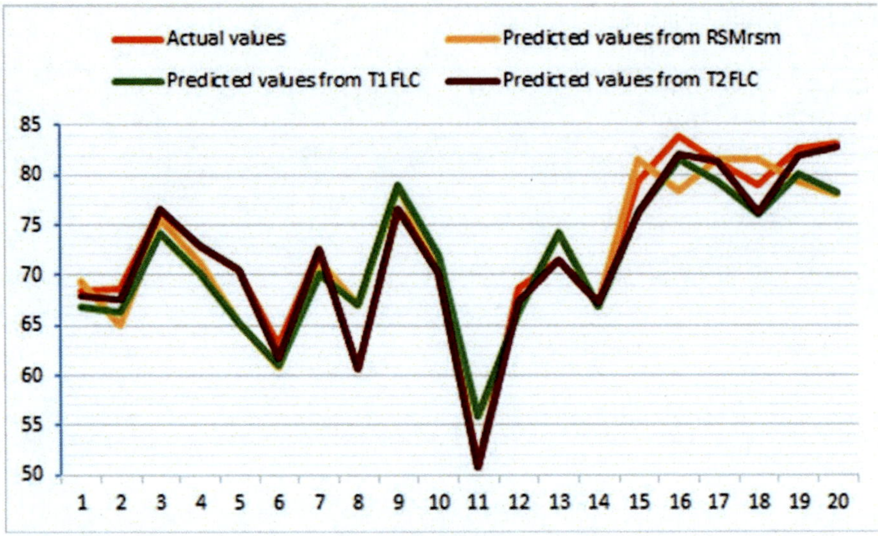

Fig. 11 Trajectory of actual value and predicted value evaluated from various soft computing techniques

Table 5 COD solubilization of POME

Data Set	Input			Output			
	Temperature	NaOH	Time	Experimental	RSM	T1FLC	T2FLC
1	30	4	24	68.48	69.4	66.81	68.01
2	30	4	72	68.74	65.05	66.28	67.6
3	30	12	24	76.51	75.6	74.19	76.51
4	30	12	72	72.9	71.24	70.2	72.9
5	40	4	24	70.55	65.29	65.29	70.55
6	40	4	72	62.98	60.92	61.01	61.72
7	40	12	24	72.61	71.49	70.1	72.61
8	40	12	72	60.67	67.12	67.12	60.67
9	36	8	48	76.59	78.89	78.89	76.59
10	43.4	8	48	71.67	70.4	71.97	70.1
11	35	1.27	48	50.7	55.82	55.82	50.7
12	35	14.73	48	68.77	66.25	66.25	67.5
13	35	8	7.64	71.37	74.27	74.27	71.37
14	35	8	88.36	67.23	66.93	66.93	67.23
15	35	8	48	79.36	81.5	76.22	76.2
16	35	8	48	83.85	78.34	81.5	82.1
17	35	8	48	81.33	81.5	79.29	81.33
18	35	8	48	79.03	81.5	76.1	76.2
19	35	8	48	82.64	79.3	80.1	81.81
20	35	8	48	83.22	78.1	78.2	82.7

this coordinated POME treatment approach as per Loh. et al. [46]. At first, the reduction of the leading wastewater stream obtains from the milling process in the palm oil plant. Secondly, in light of overseeing POME employing biogas catch and usage, preferential treatment of POME for the decrease of undesirable constituents, for example, hydrogen sulfur, in the biogas is delivered that will act as a corrosive agent for some machinery. Consequently, thermo-alkali-treatment could end up being a potential pre-treatment to improve anaerobic processing of POME and improve methane yield.

6 Conclusions

It is essential to minimize the pollution rate and to increase the solubilization % to stabilize the palm oil business model and too wide up the spectrum of its industrial profitable application. In this paper, Mamdani interval type-2 fuzzy logic inference approach which has three inputs and one output is used to predict the solubilization %. The influence of the independent parameters, reaction time, NaOH concentration, and temperature is investigated on the output parameter. The influence of the input parameter on output does not follow the trend which was traced by T2FLC techniques in an efficient way for developing the inferences train so that different sorts of procedure conditions could be anticipated. From the predicted data set (Table 4), third dimension surface plots are developed to study the variation of output concerning the interactive effects of inputs (Figs. 5, 6, and 7). From the graphical analysis, it is concluded that NaOH concentration is a more influencing parameter above all. For the validation of the performance efficiency of the T2FLC model, some of the statistical analyses like R^2, RMSE, MAPE, and MAE are evaluated and given in Table 3. In this paper, the developed model, T2FLC, is compared with other soft computing techniques such as RSM and T1FLC. The R^2 value of T2FLC is 0.991 which is the closest approach to 1 in comparison with other two, RSM and T1FLC, whose R^2 is 0.923 and 0.953, respectively. It is concluded that the prediction efficiency of the T2FLC model considers being effective and accurate. If the developed mathematical model is applied in palm oil mill effluent, it will prove to be beneficial for the engineers to set the independent parameters promptly to get the desired % of solubilization. The prediction of the % of solubilization during the effluent treatment phase will help the farm to evaluate the yield of the methane Table 5.

References

1. Atzori, G., Nissim, W.G., Caparrotta, S., Santantoni, F., Masi, E.: Seawater and water footprint in different cropping systems: A chicory (Cichorium intybus L.) case study. Agric. Water Manage. **211**, 172–177 (2019). https://doi.org/10.1016/jagwat.2018.09.040.2019

2. Sayyadi, F., Moghaddasi, R., Yazdani, S.: How climate change affects land use pattern: an Iranian provincial experience. Int J Environ Res **13**(67), 2019 (2018). https://doi.org/10.1007/s41742-018-0151-6
3. Hasanudin, U., Sugiharto, R., Haryanto, A., Setiadi, T., Fujie, K.: Palm oil mill effluent treatment and utilization to ensure the sustainability of palm oil industries. Water Sci. Technol. **72**, 1089–1095 (2015). https://doi.org/10.2166/wst.2015.311
4. Choong, Y.Y, Chou, K.W, Norli, I.: Strategies for improving biogas production of palm oil mill effluent (POME) anaerobic digestion: a critical review, Renew. Sustain. Energy Rev. **82**, 2993–3006 (2018). https://doi.org/10.1016/j.rser.201710.036
5. Iskanda, M.J, Baharum, A, Anuar, F.H, Othaman, R.: Palm oil industry in South East Asia and the effluent treatment technology-A review. Environ. Technol. Innov. **9**, 169–185 (2018). https://doi.org/10.1016/j.eti.2017.11.003
6. Angenent, L.T., Karim, K., Al-Dahhan, M.A., Wrenn, B.A., Domíguez-Espinosa, R.: Production of bioenergy and biochemicals from industrial and agricultural wastewater. Trends Biotechnol. **22**(9), 477–485 (2004). ISSN 0167-7799, https://doi.org/10.1016/j.tibtech.2004.07.001
7. Penaud, V., Delgenes, J.P., Torrijos, M., Moletta, R., Vanhoutte, B., Cans, P.: Definition of optimal conditions for the hydrolysis and acidogenesis of a pharmaceutical microbial biomass. Process Biochem. **32**(6), 515–521 (1997)
8. Kim, J.S., Park, C.H., Kim, T.H., Lee, M.G., Kim, S.Y., Kim, S.W., Lee, J.W.: Effects of various pre-treatments for enhanced anaerobic digestion with waste activated sludge. J. Biosci. Bioeng. **95**(3), 271–275 (2003)
9. Surra, E., Bernardo, M., Lapa, N., Esteves, I., Fonseca, I., Mota, J. P. : Maize cob waste pre-treatments to enhance biogas production through co-anaerobic digestion with OFMSW. Waste Manage. (2017)
10. Song, Z., Zhang, C.: Anaerobic codigestion of pretreated wheat straw with cattle manure and analysis of the microbial community. Bioresour. Technol. **186**, 128–135 (2015)
11. Valo, A., Carrere, H., Delgenes, J.P.: Thermal, chemical and thermo-chemical pre-treatment of waste activated sludge for anaerobic digestion. J. Chem. Technol. Biotechnol. **79**(11), 1197–1203 (2004)
12. Chen, H, Chang, S, Guo, Q, Hong, Y, Wu, P.: Brewery wastewater treatment using an anaerobic membrane bioreactor. Biochem. Eng. J. **105**, 321–31 (2016). https://doi.org/10.1016/j.bej.2015.10.006
13. Chaiprapat, S., Laklam, T.: Enhancing digestion efficiency of POME in anaerobic sequencing batch reactor with ozonation pretreatment and cycle time reduction. Bioresour. Technol. **102**, 4061–4068 (2011). https://doi.org/10.1016/jbiortech.2010.12.033
14. Saifuddin, N., Fazlili. S.A.: Effect of microwave and ultrasonic pretreatments on biogas production from anaerobic digestion of palm oil mill effluent. Am. J. Eng. Appl. Sci. **2**, 139–146 (2009)
15. Chou, K.W., Norli, I., Anees. A.: Evaluation of the effect of temperature, NaOH concentration and time on solubilization of palm oil mill effluent (POME) using response surface methodology (RSM). Bioresour. Technol. **101**, 22, 8616–8622 (2010). ISSN 0960-8524, https://doi.org/10.1016/j.biortech.2010.06.101
16. Appels, L., Baeyens, J., Degreve, R.: Dewil, Principles and potential of the anaerobic digestion of waste-activated sludge. Prog. Energy Combust. Sci. **34**, 755–781 (2008)
17. Aziz M.M.A., Kassim K.A., ElSergany, M., Anuar, S., Jorat, M.E., Yaacob, H., Ahsan, A., Imteaz, M.A., Arifuzzaman.: Recent advances on palm oil mill effluent (POME) pretreatment and anaerobic reactor for sustainable biogas production. Renew. Sustain. Energy Rev. **119**, 109603 (2020). ISSN 1364-0321. https://doi.org/10.1016/j.rser.2019.109603
18. Khadaroo, S.N.B.A., Poh, P.E., Gouwanda D., Grassia P.: Applicability of various pretreatment techniques to enhance the anaerobic digestion of Palm oil Mill effluent (POME): a review. J. Environ. Chem. Eng. **7**, 5 (2019). 103310, ISSN 2213-3437, https://doi.org/10.1016/j.jece.2019.103310

19. Wang, X., Zhang, L., Peng, Y., Zhang, Q., Li, J., Yang, S.: Enhancing the digestion of waste activated sludge through nitrite addition: insight on mechanism through profiles of extracellular polymeric substances (EPS) and microbial communities. J. Hazardous Mater. 369, 164–170 (2019). ISSN 0304-3894. https://doi.org/10.1016/j.jhazmat.(2019).02.023
20. Javkhlan, A., Antonio, P., Giovanni, E., Francesco, P., Piet, N.L.L.: Pretreatment methods to enhance anaerobic digestion of organic solid waste. Appl. Energy **23**, 143–156 (2014)
21. Guangxue, W.U., Zhenhu, H.U., Mark, G.H., Xinmin, Z.: Thermochemical pretreatment of meat and bone meal and its effect on methane production. Front Environ. Sci. Eng. Chin. **3**(3), 300–306 (2009)
22. Zadeh, L.: Soft computing and fuzzy logic. IEEE Softw. **11**, 48–56 (1994)
23. Shayganmehr, M., Kumar, A., Luthra, S., Garza-Reyes, J.A.: A framework for assessing sustainability in multi-tier supply chains using empirical evidence and fuzzy expert system. J. Cleaner Prod. **317**, 128302 (2021). ISSN 0959-6526, https://doi.org/10.1016/j.jclepro.2021.128302
24. Siminski, K.: FuBiNFS - fuzzy biclustering neuro-fuzzy system, Fuzzy Sets and Systems, 2021, ISSN 0165-0114, https://doi.org/10.1016/j.fss.2021.07.009
25. Zadeh, L.A.: Fuzzy sets. Inf. Control. **8**, 338–353 (1965)
26. Mamdani, E.H., Assilian, S.: An experiment in linguistic synthesis with a fuzzy logic controller. Int. J. Man-Mach. Stud. **7**, 1–13 (1975)
27. Sugeno, M.: Industrial applications of fuzzy control. North-Holland. Sole distributors for the U.S.A. and Canada, Elsevier Science Publishing company, Amsterdam, New York, NY, USA (1985)
28. Jana, D.K.: Comparative assessment on Lead removal using micellar-enhanced ultrafiltration (MEUF) based on a type-2 fuzzy logic and Response surface methodology. Sep. Purif. Technol. **207**, 28–41 (2018)
29. Jana, D.K.: Interval type-2 fuzzy logic and its application to occupational safety risk performance in industries. Soft Comput. (2017)
30. Jana, D.K.,Dey. S.: Application of fuzzy inference system to polypropylene business policy in a petrochemical plant in India. J. Cleaner Prod. **112**, 2953–2968 (2016)
31. Jankova, Z, Jana, D.K, Dostal. P: Investment decision support based on interval type-2 fuzzy expert system. Eng. Econ. **32**(2), 118–129
32. Bera, A.K., Jana, D.K, Banerjee, D., Nandy, T.: Multiple-criteria fuzzy group decision-making with multi-choice goal programming for supplier selection: a case study. Discrete Math. Algorithms Appl. **11**(03), 1950029
33. Janková, A.J., Jana, D.K.: Interval type-2 fuzzy logic expert system for investment analysis PD. Eng. Econ. ISSN: 1392-2785
34. Jana, D.K.: Novel internet of things (IOT) for controlling Indoor Temperature via Gaussian type-2 fuzzy logic. Int. J. Modell. Simul
35. Karnik, N.N., Mendel, J.M.: Introduction to type-2 fuzzy logic systems. In: Proceeding 7th International Conference on Fuzzy Systems FUZZ- IEEE, pp. 915–920, Anchorage, AK (1998)
36. Bera, A.K., Jana, D.K.: A multiple-criteria decision analysis for criticality of boiler tube failures in interval type-2 fuzzy environment. Int. J. Oper. Res. **36**(2), (2019)
37. Dey, S., Jana, D.K., Khatua, P.K., Mukherjee, A.: Application of fuzzy inference techniques in the production of eco-friendly aminoplast based modified resins for plywood panel industries. Int. J. Fuzzy Comput. Modell. **2**(4), 303–321
38. Mukherjee, A., Roy, K., Jana, D.K., Hossain, S.A.: Qualitative model optimization of almond (Terminalia catappa) oil using soxhlet extraction in type-2 fuzzy environment. Soft Comput. **24**(1), 41–51
39. Bera, A.K., Jana, D.K., Banerjee, D., Nandy, T.: A group evaluation method for supplier selection based on GSCM practices in an Indian manufacturing company. In: International Conference on Information Technology and Applied Mathematics
40. Sahoo, P., Jana, D.K., Panigrahi, G.: Interval type-2 Fuzzy logic and its application to profit maximization solid transportation problem in mustard oil industry. In: International Conference on Information Technology and Applied Mathematics

41. Jana, D.K., Das, B., Maiti, M.: Multi-item partial backlogging inventory models over random planninghorizon in random fuzzy environment. Appl. Soft Comput. **21**, 12–27 (2014)
42. Castillo, O., Melin, P.: Type-2 Fuzzy Logic: Theory and Applications, Studies in Fuzziness and Soft Computing, Springer, Berlin, p. 223 (2008)
43. APHA, AWWA, WPCF: Standard Methods for the Examination of Water and Wastewater. American Public Health Association (2005)
44. Lin, J.G., Ma, Y.S., Allen Chao, C., Huang, C.L.: BMP test on chemically pretreated sludge. Bioresour. Technol. **68**(2), 187–192 (1999)
45. Heo, N.H., Park, S.C., Kang, H.: Solubilization of waste activated sludge by alkaline pretreatment and biochemical methane potential (BMP) tests for anaerobic co-digestion of municipal organic waste. Water Sci. Technol. **48**(8), 211–219 (2003)
46. Loh, S.K., Nasrin, A.B., Azri, S.M, Adela, B.N., Muzzammil, N., Jay, T.D., Eleanor, R.A.S., Lim, W.S., Choo, Y.M., Kaltschmitt, M.: First Report on Malaysia's experiences and development in biogas capture and utilization from palm oil mill effluent under the Economic Transformation Programme: Current and future perspectives. Renew. Sustain. Energy Rev. **74**, 1257–1274, 1364-0321, (2017) https://doi.org/10.1016/j.rser.2017.02.066